Preface

This book is intended primarily for bacteriologists and bacteriological technicians working in routine medical diagnostic bacteriology laboratories, but it is also likely to be of interest to similar people working in other types of microbiology laboratories.

Brief expositions of many aids and the applications to which they have been put hitherto are given, but this is *not* a comprehensive review of all available devices in all categories. Such a review would be an enormous tome, since one could write a dissertation on each category of device included, and be out of date before it was published. I merely give a few examples. The suggestion is that the reader should ask himself if he has an application for any of the devices discussed, or any similar devices, in his own laboratory.

It is not appropriate to discuss the philosophy of introducing mechanization into microbiology in this book. (Care must be taken not to confuse mechanization with automation, which, at its simplest, involves far more sophisticated techniques and apparatus than any presently in regular use in diagnostic microbiology.) A number of questions, such as whether it is better to use equipment designed to perform one specific procedure or more flexible equipment and whether the sole or primary purpose of mechanization should be cost effectiveness, have to be considered. My thoughts on many such matters were published in 1973* and are unchanged. But one may well have very special reasons exclusive to one's laboratory for introducing a particular type of machine, and should not necessarily be put off by the fact that the conditions pertaining do not satisfy the so-called normally accepted criteria, about all of which there is still considerable controversy.

If the reader decides he has an application for a technological aid, it is up to him to study the literature, to look carefully at all possibilities and then to go out and acquire the apparatus he requires and develop his own procedure. Incidentally, that is the time to consider updating one's technique and reorganising one's laboratory, as opposed to merely mechanizing an existing technique.

There is no definitive section on microbiological safety, but one should always be cautious, particularly because, as I indicate in the text, some inventions seem to me to be positively dangerous. Even when that is not the case, there is often no hint that the safety of the device was considered. There appears to be a serious lack of awareness of microbiological safety by both inventor and user. No general rules can be made and certainly no regime for the safety testing of microbiological apparatus has been universally adopted. One has to devise one's own scheme, but it is very important to do so. Much more awareness of the potential hazards of devices used in microbiological laboratories is needed.

No doubt readers will be able to point out categories of devices to which no reference has been made but which might have been included. One has, of course, to make a positive decision about such matters, since neither time nor

* Trotman, R. E. (1973). The philosophy of the application of automatic methods to hospital diagnostic bacteriology. *Bio-Med. Engng* **8**, 519.

space are unlimited. I have included those categories of device about which I feel competent to write and from the use of which in my view the greatest benefit can be derived: the scope of the book has been restricted to relatively simple aids, because very little work leading to the development of more complex technological systems has produced practical systems of use to more than a handful of very specialized laboratories.

It will be appreciated that one requires a great deal of assistance in producing a volume such as this. In particular, authors of published papers and books have given invaluable assistance in supplying many photographs, sketches, drawings and graphs relevant to their devices. I am most grateful to them all for so readily putting themselves to so much trouble for me and for giving permission to use the material in this way. I wish also to express my gratitude to the following: the editors of journals and publishers of books for permission to reproduce photographs and drawings that have previously appeared in their publications; the St Mary's Hospital Medical School's Audio-Visual Communications Department for producing many of the photographs (some of the tasks I presented it with, particularly in producing Fig. 1, were very difficult, but I am sure readers will agree that the effort was well worth while); Mr Kevin Byrne for producing many of the drawings; and Mrs Heather Downey and Mrs Peggy Howard for typing the manuscript.

R.E.T.

London, 1977.

Contents

I
Devices for distributing infected and/or sterile fluids

There is clearly a need for devices for distributing fluids, such as antibiotic solutions and nutrient broth, aseptically and for distributing infected fluids. We now discuss the most commonly used.

Pipetting small volumes of fluids

Dropping pipettes

In 1913 Donald described a device for producing measured small volumes of liquids, and he gave more details of its performance in 1915. The apparatus is a glass pipette with a perfectly cylindrical capillary, from which a fixed number of drops gives 1 ml of liquid. The actual number depends on the outside diameter of the capillary, and Donald described those devices giving approximately 130 drops/ml to 10 drops/ml. Wilson (1922, 1935) and Miles and Misra (1938) have exhaustively investigated various applications of the pipettes.

A pipette in common use today is one which gives 50 drops/ml; it is known as the 50 dropper. Donald showed that if the difference in outside diameter varies from 1.016 to 0.914 mm the number of drops/ml varies from 50 to 54, which is an error of $\pm 4\%$. This doubtless applies only if the capillary is cylindrical and if the tip of the pipette is smooth and cut at right angles to the cylinder.

There is no reason to doubt that, when carefully manufactured, 50 droppers are as accurate as this, but it is most unlikely that they are manufactured sufficiently accurately under modern routine laboratory conditions. We found that the mean volume of 12 drops dispensed from a 50 dropper manufactured in our laboratory was 0.02 ml but that the volume of individual drops varied between 0.0237 and 0.0175 ml. We found also that the mean volume of 12 drops from each of 12 different pipettes varied between 0.0172 and 0.0244 ml. These results are hardly surprising when one considers the difficulty in keeping the capillary cylindrical during manufacture, and also the difficulty in cutting the tip smoothly to within 0.1 mm in diameter, which is necessary for an accuracy of $\pm 4\%$.

One may argue that one should pay more attention to the manufacturing process, but this is a very time-consuming, and therefore costly, procedure. It is not surprising, therefore, that many people have devoted considerable efforts to devising more reliable devices for distributing small volumes of fluids.

Today a variety of dropping pipettes, both disposable and reusable, are commercially available and details of their evaluations have been published.

Unfortunately, these are usually hidden away in papers the primary purpose of which is to describe a technique in which the use of the pipettes is incidental, in such papers as those describing serial dilution techniques for antibiotic sensitivity testing.

MacLowry et al. (1970) used some 50 µl dropping pipettes marketed by Cooke Engineering Company* to dispense diluent, broth and test organisms. These pipettes are calibrated to deliver 50.0 µl of 0.9% NaCl/drop, but the size of drop varies according to the surface tension of the fluid being dispensed. The authors state, for example, that the pipettes dispensed 48.8 µl of human serum, 41.1 µl of Mueller-Hinton broth and 40.8 µl of Trypticase Soy broth, but they did not give evidence of having established these figures experimentally.

The particular pipettes used are reusable. They were placed in Amphyl for between 2 and 6 h, rinsed three to four times in distilled water, inverted and allowed to dry overnight. Each pipette was then wrapped in a paper towel sterilized with ethylene oxide and allowed to remain at room temperature for 1–2 h before reuse.

The pipettes were recalibrated every six months, or more frequently if there was evidence to suggest the results of the procedure were poor. The authors referred to a commercially available pneumatically operated dispensing device, but gave no details.

Tilton et al. (1973) used 50 µl disposable pipettes to distribute inoculum in an antibiotic sensitivity test procedure. To test the accuracy and repeatability of the pipettes they selected samples of pipettes at random, delivered 20 drops from each and determined the average drop volume for each of 0.85% saline, brain-heart infusion broth, Mueller-Hinton broth and Trypticase Soy broth. Each experiment was repeated ten times and they found the mean drop volume to be 49.2 µl, 47.0 µl, 46.0 µl and 46.2 µl respectively.

Gavan and Town (1970) found that the average size of drops dispensed by 50 µl dropper pipettes varies depending on whether or not the plastic microwell plate into which the drops are dispensed is charged with static electricity. They found that an average volume of 39.0 µl ± 10 µl was dispensed if the plate was charged and an average volume of 47.4 µl ± 3.6 µl was dispensed if it was not; these figures are for dispensing distilled water. Because of this phenomenon Tilton et al. (1973) held their dropping pipettes 2 cm above the microwell plate to keep errors due to static electricity to a minimum.

Fitzgerald et al. (1974) briefly referred to 10 µl pipette droppers. They did not report their evaluation of them but merely indicated that they required the maintenance of a constant pressure to deliver the appropriate amount of reagent without introducing air bubbles.

The above brief references to some evaluations of commercially available dropping pipettes, which are very widely used, show that they are most valuable but that, as with so many devices, a carefully controlled routine for their use and their care has to be established, and that constant checking of their performance is necessary. These factors are often neglected in such

* Cooke Engineering Company, now incorporated in Dynatech Laboratories Inc., 900 Slaters Lane, Alexandria, Virginia 22314, USA.

simple devices, because one tends to take the attitude that there is nothing to go wrong with them. The fact is that their performance very easily deteriorates and it should never be taken for granted.

Semi-automatic pipettes

In addition to care, some skill is required in using dropping pipettes. In consequence, in recent years a wide variety of semi-automatic high-precision microlitre pipettes have been devised. Although they vary in detail, all consist, essentially, of a high-precision syringe with a spring-loaded plunger which has either a fixed stroke length, as in the Eppendorf* type, or a variable stroke length, as in one Pipetman† type. The plunger is drawn up and down to suck up and discharge a volume of fluid, which volume depends on the stroke length in the conventional way. A disposable presterilized tip is attached to the bottom of the barrel of the syringe, and one should not attempt to use these devices without the tip. In normal usage, the fluid to be dispensed comes into contact only with the tip, which for the highest accuracy of dispensing should be held in contact with the side of the vessel into which the fluid is discharged. In most types, the syringe plunger may be moved marginally beyond the normal end of stroke when expelling fluid to ensure that the last drop of it is ejected from the tip. A photograph showing the principle of these pipettes is shown in Fig. 1.

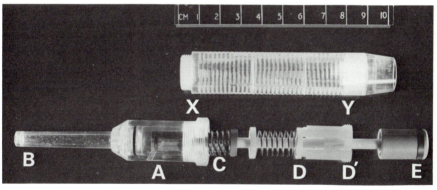

Fig. 1. Exploded view of an Eppendorf pipette. A: syringe barrel. B: rod to which the disposable tip is attached. C: spring-loaded syringe plunger. D, D': stops, the distance between which determines the stroke length of the plunger. E: actuating rod. X, Y: the threaded parts of the pipette are screwed into the corresponding threaded parts of the barrel. (Loan of exhibition model of the pipette courtesy of Anderman Co. Ltd.)

Volumes dispensed vary, but are usually a maximum of 5 ml. For example, at the time of writing, Eppendorf manufacture 25 pipettes in the range of 5 µl to 1 ml and Gilson† manufacture four Pipetman types in which the volume dispensed is continuously variable between 0 and 20 µl, 0 and 200 µl, 0 and 1 ml, and 0 and 5 ml respectively.

---

* Eppendorfer Gerätebau, Netheler & Hinz GmbH, 2000 Hamburg 63, Postfach 630324, West Germany.
† Gilson France, 69/72 Rue Gambetta, 95400 Villiers le Bel, France.

Joyce and Tyler (1973) evaluated the performance of five different types of semi-automatic disposable tip pipettes. The models tested and the volume dispensed are set out in Table 1.

TABLE 1

Type of pipette	Volume dispensed (μl)
Biopipette	100–200 (preset)
Biopipette	100–1,000 (100 multiples)
New Eppendorf	100 (preset)
Finnpipette	200–1,000 (adjustable)
Gilson Pipetman	0–200 (adjustable)
Oxford Sampler	100 (preset)
Oxford Sampler	200 (preset)
Marburg Eppendorf*	100 (preset)
Marburg Eppendorf*	1,000 (preset)

*The last two pipettes are no longer marketed.

They found that in most cases the volume of deionized distilled water dispensed was temperature dependent, there was a tendency for the pipettes to dispense progressively smaller volumes until the temperature reached 30 to 32 deg C, which, they state, is approximately the closed-fist hand temperature of their laboratory staff. In one case they found the temperature-dependent effect to be 'alarming'. Unfortunately, they did not state the accuracy of dispensing when the temperature has stabilized at the hand temperature, but they did say that when there was a marked change in volume dispensed this usually stabilized by the time 12 successive samples had been ejected. They also found that two continuously adjustable pipettes, Gilson Pipetman and Finnpipette,* were as precise as those (such as the Eppendorf pipettes) that deliver a fixed volume. They believed the adjustable pipettes to be generally preferable to the fixed-volume pipettes. Of the fixed-volume pipettes there were two makes in particular, the Eppendorf's and the Oxford's,† in which it was advisable to prewarm the pipette for at least 8 min before use, if highly repeatable samples were required.

Ellis (1973) evaluated the Eppendorf pipettes and the Oxford samplers and he also found that the volume delivered was temperature dependent. He indicated that the conditions have to be well controlled for the performance of the pipettes to comply with the performance quoted by the manufacturers.

He calibrated the apparatus by weighing the amount of water or solution dispensed each delivery. The mean weight and standard deviation of the mean were determined from at least 6 weighings, and the volumes were calculated using the density of the water or solution dispensed. Using an

* Labsystems OY, Pulttitie 9, 00810 Helsinki, Finland.
† Oxford Laboratories (International), 1149 Chess Drive, Foster City, California, USA.

Eppendorf 0.05 ml pipette, he found that the volume delivered at 25 deg C and at 0 deg C was 0.0501 ml ± 0.0002 ml and 0.0467 ml ± 0.0005 ml respectively. Experiments to determine the effects of prerinsing the tip, of the repetitive usage of tips and of the variation of the temperature over the whole pipette unit were also performed, but unfortunately, the evidence given was concerned mainly with the temperature variation effect. It was found that all parts of the system should be at the same temperature. For example, if the temperature of the syringe differed from that of the fluid being dispensed, the volume delivered was very dependent on the exact procedure followed.

Robinson and Johnson (1974) assessed the accuracy of the smaller volume (5–100 µl) semi-automatic pipettes. They found that the precision of dispensing decreased as the volume decreased. For example, using an adjustable Finnpipette in the range of 5–50 µl they found that when set to deliver 50 µl the mean of 10 volumes dispensed was 50.84 µl, giving an error of + 1.6%, whereas when the same pipette was set to deliver 5 µl the mean of 10 volumes was 5.49 µl, which is an error of +9%. The coefficient of variation was found to be 0.4% and 2.4% respectively. They also found that with a 50–250 µl Finnpipette set to deliver 50 µl, the mean of 10 volumes dispensed was 52.06 µl, an error of + 4.12%. They felt that these results suggested that the errors were related to the stroke length of the plunger of the syringe. This may well be so, although it is probably also related to the fact that the plunger is not directly in contact with the fluid being sucked up and dispensed; there is an air gap in between which is, of course, highly compressible, and the longer the stroke length the less significant errors due to this effect are likely to be.

Examples of the results obtained with fixed-volume Eppendorf pipettes are: a 5 µl pipette delivered a mean volume (of 10 volumes) of 5.32 µl, an error of + 6.4%, and a 10 µl pipette delivered a mean volume of 10.1 µl, an error of + 1%.

An interesting observation was that although Robinson and Johnson had been using their pipettes for a period of up to two years, they did not find any progressive deterioration in the performance over that period. Furthermore, they attributed bad performance to spurious contamination in the capillary of the pipette, to wear and tear and to lack of lubrication of the plunger mechanism. They found that on routine servicing the performance deteriorated, but this was believed to be due to contamination in the capillary which, in turn, was believed to be due to reverse pipetting, in which some tips are filled almost to capacity; this clearly increases the likelihood of contamination. In summary, Robinson and Johnson found that when dispensing volumes of 50 µl and above, the pipettes conformed to the manufacturer's claims, but that when dispensing smaller volumes there was a progressively greater discrepancy between performance and the manufacturer's claims.

Bousfield et al. (1973) used a 100 µl Eppendorf pipette to dispense cell suspension in a method for counting bacteria: 100 µl volumes of cell suspension were pipetted into 100 ml portions of diluent and mixed. Using the Eppendorf pipette, 100 µl of this dilution was transferred to a further 100 ml of diluent, thus giving a final dilution of 10^6; 100 µl of this dilution was then deposited as a series of 5 or 6 drops on the surface of a nutrient agar plate. The total number of colonies in the drops were counted after incubation overnight. The original suspension contained 2×10^9 *Escherichia coli*/ml in a mineral salt solution, so each drop contained approximately 200 cells. Unfortunately, they did not analyse the performance of the Eppendorf pipette separately.

Bousfield and his colleagues compared this method with the pour plate and the spread plate methods. They found it slightly less precise than the pour plate method but more precise than the spread plate method. However, it involves less preparative work and is easier and quicker to perform than the pour plate method.

From the above discussion on semi-automatic disposable tip pipettes one can understand why they have begun to supersede the old dropping pipette. Nevertheless, all is not plan sailing and I can do no better than quote Joyce and Tyler (1973). 'However, they [the disposable tip pipettes] can differ considerably in ease of use, price, accuracy and precision.' So, like the dropping pipette, their performance cannot be taken for granted. Furthermore, they have one disadvantage in that they will dispense 1 drop of fluid per filling whereas a dropping pipette will dispense many drops per filling. We return to that matter later.

Recent additions to the pipettes that are commercially available are the multichannel Labpipettes.* These are 4– or 8–channel pipettes very similar in principle and in operation to the single-channel Finnpipette. There are 22 pipettes, in the range 5–200 µl, available in both the 4– and 8–channel systems. The 'average accuracy' is claimed to be $\pm 1.5\%$. The tips are suitable for both single– and multichannel versions.

Transferring small volumes of fluids

Diluting loops

In 1955 Takatsy devised a simple device for titrating small measured volumes of liquid, which has had considerable impact in a variety of pathological laboratories. This device, known as the microtitrator loop, consisted of a lightly wound wire spiral closed at each end (Fig. 2). One of his loops titrated

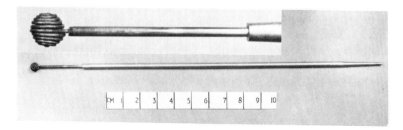

Fig. 2. A modern wire spiral-type microtitrator loop.

*Labsystems OY, Pulttitie 9, 00810 Helsinki, Finland.

0.025 ml of liquid and another titrated 0.05 ml. One simply dipped the loop into the liquid, rotated the loop a few times, removed it from the liquid and the appropriate volume was held in the wire spiral by capillary action. The liquid so held can be transferred and mixed with another liquid.

The microtitrator loop was first used (for serological titrations) by Sever (1962), but he found many difficulties. The wire, which was not made of stainless steel, tended to rust and it was easily distorted, thus becoming very inaccurate. Furthermore, in order to ensure that it transferred the correct volume, the loop had to be prewet and all particles and grease had to be removed before use. Those factors indicate that the original loops were very delicate and became inaccurate very easily.

In recent years, the quality of the microtitrator (now more commonly called the microdilutor) loops, many of which are no longer lightly wound wire spirals but are made from solid metal (Fig. 3), has improved considerably and many are commercially available.

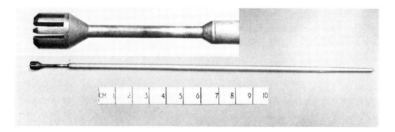

Fig. 3. The type of microtitrator loop which is fabricated from solid metal.

A detailed comparison of the performance of both types of loop was carried out by Ashcroft et al. (1971). This work was concerned primarily with the performance of the loops in the haemagglutination inhibition test, so we will leave detailed discussion of this paper until later (see p. 59). Suffice it to say here that Ashcroft and his co-workers found no really significant difference in the performance of the two types of loop. They found the loops transferred, on average, 90% of the nominal volume of fluid. They found also that the performance did not deteriorate significantly over a 'period of use'. Unfortunately, they did not say how long this period was.

Microtitre techniques are becoming increasingly widely used, primarily because of the saving in reagents that may be effected by using them; earlier techniques use much larger volumes of reagents.

A very common application of the microdilutor loops is in serial dilution antibiotic sensitivity testing. Marymont and Wentz (1966) used a 50 µl loop for this purpose. They used 50 µl volumes of diluent and, with an initial antibiotic concentration of 200 µg/ml, serially diluted the antibiotic 12 times, using the now familiar multiwell reusable or disposable plastic plates. The minimum concentrations of up to three antibiotics required to inhibit growth of up to 60 bacteria were determined by this method and also by the standard tube dilution method; a total of 111 minimum inhibitory concentrations (MICs) were measured by each method. In 70 cases the tube dilution method gave a MIC which was either 1 or 2 dilutions higher than that given by the microtitre method, and in 4 cases the tube dilution method gave a minimum inhibitory concentration which was 3 or more dilutions higher than that given by the microtitre method.

Harwick et al. (1968) also used the microtitrator system to determine the MIC of antibiotics, and, in addition, to determine the minimum bactericidal concentration (MBC). Seventy-six organisms isolated from blood cultures were tested against up to 16 antibiotics (penicillin G, kanamycin, ampicillin, methicillin, tetracycline, chloramphenicol, lincomycin, 5-cloro-7-deoxylincomycin, carbenicillin, streptomycin, polymyxin B, colistin sulphate, rifampicin, cephalothin, erythromycin glucoheptate and vancomycin) by both the standard tube dilution method and by the micro method.

The standard tube dilution method was as follows. Suspensions of antibiotics in distilled water, at a concentration of 1,000 µl/ml, were prepared and serially diluted 9 times; the dilutions were twofold; 0.1 ml of each dilution of antibiotic solution was added to a test tube containing 0.8 ml of nutrient broth and the tubes were frozen at -16 deg C and stored. As required, tubes were thawed and 0.1 ml of culture containing 10^5 organisms/ml was added. Cultures were incubated for 18 h at 37 deg C and the first dilution with no apparent growth was considered to be the MIC. This culture and all cultures in subsequent dilution tubes were subcultivated on to blood agar and incubated for a further 24 h. The original cultures were also incubated for a further 24 h. Colonies growing on each subculture were studied and the lowest dilution from which one or no colonies grew was regarded as the MBC (a 99.9% kill).

In the microtitre method, cultures were prepared as before but antibiotic solutions of a concentration of 800 µg/ml were used. The technique was as follows: 50 µl of nutrient broth was added to each well (U bottom) of an 8 × 12 well plate; 50 µl of antibiotic stock solution was added to each well in column 2 and the fluid in the wells mixed by rotating the microdilutors for approximately 5 s. They were then transferred to the wells in column 3 and rotated to mix the broth and antibiotic. This process was continued so that serial dilutions of the original antibiotic solution were made from column 3 through to column 7. Then 50 µl of nutrient broth was added to each well in column 8 and 0.1 ml of nutrient broth was added to each well in the plate. The plates were stored at -16 deg C. To perform the test, 50 µl of the culture was added to each well in columns 1–7, the final titre being approximately 10^5 organisms/ml. Column 8 was a media control. The plates were incubated for 18 h. The minimum inhibitory concentrations and the minimum bactericidal concentrations were determined as before.

Minimum inhibitory concentrations determined by the macro and micro methods were compared in 922 determinations and 853 (92.5%) agreed to within 2 dilutions. Minimum bactericidal concentrations were compared in 903 determinations and 807 (87.4%) agreed to within 2 dilutions.

The precise combinations of organism/antibiotic tested were not given, but the way in which each of the 16 antibiotics performed was given. The results ranged between 100% agreement (45 tests) with carbenicillin and 75%

agreement (18 out of 24 tests) with lincomycin for the MICs, and between 100% agreement (24 tests) with methicillin and 81% agreement (57 out of 70 tests) with polymyxin B for the MBCs.

The authors found that the microtitration techniques produced a good saving in the labour costs of performing the determinations, and that a technician can perform three times as many determinations with the micro method as he can perform in the same time with the standard tube dilution method. They estimated the total cost to be about one-third of the standard method.

Harwick and his colleagues also performed reproducibility tests on the microtitration system. Isolates of *Escherichia coli, Klebsiella pneumoniae, Proteus mirabilis, Staphylococcus aureus* and *Streptococcus faecalis* were each tested 11 times against all 16 antibiotics. Their results indicated that a 95–98% probability existed that the MIC end-point observed would fall within 1 dilution of the mode, and that a 93–97% probability existed that the MBC end-point would fall within 1 dilution of the mode.

Semi-automatic diluting apparatus

Although it is not specifically stated, one is left to conclude that in the methods described above, the loops were held and manipulated manually. Quite clearly, when one is trying to hold and manipulate a number of loops simultaneously, a considerable amount of manual dexterity is required. In consequence, a variety of semi-automatic devices, in which the loops are supported and rotated automatically, have been devised and one of the earliest was by MacLowry and Marsh (1968).

In their device (Fig. 4), which was devised specifically for serial dilution antibiotic sensitivity testing, 12 microdilutors, of either 25 or 50 µl capacity, are attached in a straight line to a vertically movable head; all dilutors may be rotated simultaneously, a preset number of times, by means of an electric motor. The microtitre (U) plates in which the tests are carried out, and the reservoirs of reagents, are placed on a holder, which moves horizontally under the microdilutor loops: the spacing between the loops is identical to the spacing between the wells in the plates. The horizontal movement of the holder and the vertical movement of the head in which the loops are supported are controlled manually by means of levers. Actuation of the motor controlling rotation of the loops is performed automatically by the head as it is lowered.

They gave little detail of their experiments, but they measured minimum inhibitory concentrations of a variety of combinations of organisms and antibiotics by their semi-automatic microtitre apparatus and compared the results with the results obtained by the standard tube dilution method. Eighty-eight MICs were performed by both methods: 68% produced the same end-point, 19% produced end-points which varied by 1 dilution, 7% by 2 dilutions and the remaining 6% by 3 dilutions. The differences did not appear to be related to any particular antibiotic or organism. They also carried out reproducibility studies on their equipment, and of 336 repetitive assays, 86% gave identical results, 13% differed by 1 dilution and 1% by 2 dilutions.

(a)

(b)

Fig. 4. The machine devised by MacLowry and Marsh for serial dilution antibiotic sensitivity testing by the microtitre method. (a) The complete apparatus. The reservoirs in the foreground, immediately in front of the microtitre tray, contain the antibiotic solutions. The loops are shown situated over a reservoir containing a solution in which they are rinsed at the end of the cycle. (b) Close-up view of the microdilutors, so arranged to utilize only 8 of the 12 loops. A: reservoirs containing antibiotic solutions. R: reservoir of solution in which the loops are rinsed. (From MacLowry, James D. and Marsh, Harry H. (1968). Semiautomatic microtechnique for serial dilution antibiotic sensitivity testing in the clinical laboratory. *J. Lab. clin. Med.* **72**, 685–687.) (Photographs courtesy of Dr James D. MacLowry.)

MacLowry and his colleagues (1970) later published a most valuable paper in which considerable detail of their methodology was given. They discussed the equipment, the calibration and sterilization procedures, the media, the preparation of the antimicrobial agents, and the quality control procedures they introduced. In addition, some most useful tips on reading the plates were also given.

They evaluated the 50 μl microdilutors independently, which were found to deliver a mean volume of 50.4 μl (standard deviation 0.65) of 0.9% NaCl. After 8 months' use the mean volume was found to be 51.6 μl (standard deviation 0.64). Mean volumes of 49.4 and 49.9 μl (with standard deviations 0.36 and 0.71 respectively) of Trypticase Soy broth (and also of serum) were delivered immediately and after 8 months' use respectively.

The preparation of the microdilutors was as follows. They were flamed to incandescence, quenched in distilled water, touched with blotting paper to remove most of the water and then flamed briefly to dryness. This procedure was adopted every morning.

In the experiments described below a commercial version (American Instrument Company*) of the semi-automatic instrument referred to in their earlier paper was used. No details of the machine were given other than a statement to the effect that the instrument holds twelve 50 μl microdilutors.

The accuracy of the twofold dilutions was determined by diluting a standard solution of sodium chloride and measuring the chloride content in each well. The eighth dilution was found to be $\pm 5\%$ of the expected value. All previous dilutions were found to be more accurate than that. The number of experiments used to determine these values was not given. Of course, the error in the serial dilution process is due to the errors in pipetting the diluent and other solutions (discussed on p. 3), as well as due to the serial diluting process. It is not at all clear if the expected values of sodium chloride concentration quoted took that into account or not.

Their apparatus was normally kept in an 'ultraviolet (uv) hood', except when dilutions were being made; the uv light was turned on if the equipment was left unused for more than 15 minutes. The apparatus had been operated for periods of up to 6 h duration without the uv light, but under the hood. The authors stated there had been no contamination problem.

The plates were prepared by dispensing 50 μl of broth diluent into all wells and 50 μl of the initial concentration of each antibiotic into the first well of each row. As the dilutions were made the operator noted the height of the diluent in each of the wells to ensure that it did not rise above the top of the microdilutor 'chamber'; it is not clear if this means the well or the head of the dilutor. The height to which the diluent rises in the chamber should be constant. After use the loops were rotated in sterile distilled water, blotted on filter paper and flamed briefly to dryness. Then 50 μl of a 1/1,000 dilution of a suspension of the organism, prepared by mixing 6–10 isolated colonies in 2 ml of broth, was added to each well. In the case of *Proteus* organisms, a 1/10,000 dilution was used, since the end-points were said to be much sharper and much more reproducible with the smaller inoculum.

*American Instrument Company, 8030 Georgia Avenue, Silver Spring, Maryland 20910, USA.

Twelve antibiotics were used at one time, but no positive or negative controls were included. This was because there was usually growth at some concentration of antibiotic, this culture serving as a positive control, and there was usually no growth at a higher concentration of antibiotic, this culture serving as a negative control. The plates were incubated at 37 deg C for approximately 18 h, but in some (unspecified) organisms a preliminary reading could be made after 4–6 h. In such cases, it was found that the final reading taken after 18 h did not usually differ from the preliminary reading by more than 2 dilutions.

The plates were read by placing them over a plastic sheet lined with black strips so as to produce a dark background. Bacterial growth in the system was usually seen as a button of bacteria which had settled to the bottom of the U-shaped well (they did not use V-shaped wells because the small plastic button at the bottom of the well made reading difficult), but some organisms which grow as very small granules were difficult to see; for example, staphylococci with erythromycin and many organisms with tetracycline and chloramphenicol were difficult to interpret because of the small amount of growth which occurred in many of the wells. For this type of growth the end-point was arbitrarily designated as the lowest concentration in which the button of bacteria was less than 0.5 mm diameter. MacLowry and his co-workers found that if this criterion was used, reproducible results were obtained by different observers.

They stressed the need for the careful control of the potency of the antibiotics and, in particular, the careful calibration of the microdilutors (and also of the pipettes used to distribute the diluent). Recalibration of the dilutors was performed every 6 months or more frequently if differences in the results could not otherwise be accounted for.

Quality control organisms were used for assaying the potency of the antibiotic. These were, usually, *Escherichia coli* and *Staphylococcus aureus*. The *E. coli* was tested with those antibiotics used for gram-negative organisms and *Staph. aureus* with those used for gram-positive organisms.

A very similar procedure to that used to determine the MIC of organisms was used to determine serum antibiotic levels and also the level of antibiotic in other body fluids, such as urine and CSF. Brief details of the techniques were given.

The authors stated that in their laboratories three technologists performed 8,000–10,000 serial dilution tests per month. They found that the technique is neither simpler nor faster than the multodisc method for determining antibiotic sensitivities, but that it is of the order of 10 times faster than the standard tube dilution method, which is, of course, usually performed manually.

Chitwood (1969) used a simple handle which holds up to 12 loops for performing the dilutions required in antibiotic sensitivity testing; he used 25 µl loops. No other details of the handle were given. Isolates of a variety of gram-negative rods, such as *Pseudomonas, Salmonella, Shigella, Klebsiella, Enterobacter* and *Escherichia*, and of gram-positive organisms, such as *Staphylococcus*, were tested against a variety of antibiotics, in particular, cephalothin, ampicillin, colistin, neomycin, tetracycline, kanamycin, polymyxin B, streptomycin, chloramphenicol, gentamicin, cephaloridine, lincomycin, erythromycin, penicillin G, carbenicillin and cloxacillin. Results obtained by the micro method were compared with the results obtained by the standard tube dilution method.

They gave a useful table in which they showed how the correlation between the two methods improved as the operator became more familiar with the procedure. Initially, 55 out of 71 determinations (77%) agreed to within ± 1 tube dilution (which is assumed to be acceptable). After a short period of familiarization the correlation improved and after the technique had been in use for 5 months 94% agreement (168 out of 179 determinations) was achieved.

The authors found that the most time-consuming procedure when this handle was used was the addition of diluent to each cup, for which they used a calibrated pipette dropper. MacLowry et al. (1970) found that this handle could give an accuracy comparable to that obtained with their semi-automatic machine, but that it was slower in use when a large number of plates was processed.

Chitwood sterilized his loops by heat, whereas Gavan and Town (1970), who also used the same handle for holding the loops, sterilized theirs (which were 50 μl loops) by dipping them in acetone and subsequently passing them through a flame; they were then air cooled.

They found that in fact the loops transferred a different volume to the 50 μl nominal volume. The mean of 8 tests gave a volume transfer of 53.4 μl; the range was from 50.1 to 55.6 μl. (They also found, as reported on p. 2, that the dropper pipettes used to deliver the diluent delivered a volume which depended on whether or not the multiwell plates used for these tests were charged with static electricity. The mean of 11 volumes discharged into a plate that was charged with electricity was 39.0 μl—the range was from 33.0 to 46.1 μl—whereas the mean of 11 volumes discharged into a plate which was not charged with static electricity was 47.4 μl—the range was from 44.0 to 49.3 μl.)

The results of measuring the MIC in plates from which static electricity had not been removed when compared with the results obtained by the standard tube dilution method was given for the following antibiotic/organism combinations: kanamycin/*Enterobacter*, gentamicin/*Pseudomonas*, vancomycin/*Enterococcus*, gentamicin/*Bacillus subtilis*. The micro method gave a result invariably 2 dilutions lower than the result obtained by the standard tube dilution method. This was assumed to be due to the static electricity on the plastic plates, but whether this was solely the result of errors in dispensing diluent due to this phenomenon was not investigated.

Gavan and Town compared the results obtained by the standard tube dilution method with the results obtained by the micro method using plates with static electricity and with those obtained by the micro method using plates which had first been exposed to ultraviolet radiation, for between 2 and 5 h, to discharge the electricity. The test for each combination of antibiotic/organism was repeated five times by the standard tube dilution method and seven times by each of the micro test procedures. Using the paired sample t-test, it was found that the mean of the MIC values obtained by the micro method using unexposed plates differed significantly from the values obtained by the standard tube dilution method, but that differences in MIC values obtained with the exposed microplates and values obtained by the standard tube method were not significant.

Reproducibility of the microdilution technique was investigated; 300 replicates were observed and it was found that 94.8% of the tests were within ±1 dilution from the median. Similarly, they found that in 105 replicate tests by the macrodilution technique, 99.1% were within ±1 dilution.

The authors concluded that if the operator is aware of the difficulties involved in the microdilution method and if the static electricity is dissipated from the plates, the method can give reliable results which compare favourably with those obtained with the conventional tube dilution method.

More sophisticated diluting apparatus

At the time MacLowry and Marsh described their semi-automatic diluting device, Goss and Cimijotti (1968) described the use of a slightly more sophisticated machine, called the Autotitre. This machine automatically delivers diluent to the wells, makes up to 14 twofold serial dilutions, and then adds inoculated broth to each well.

In this apparatus (Fig. 5) a set of eight 50 µl microtitre dilution loops is supported on a carriage and is lowered into a tank containing 70% ethyl alcohol; the carriage supporting the loops then moves to a position in which the loops are blotted. The carriage is moved so that the loops are over the first row of wells, in a multiwell tray, in which antibiotics are contained; the loops are lowered to take up 50 µl. Simultaneously, 50 µl of diluent is delivered to each well in the second row of wells, by means of 8 syringes mounted in front of the loops on the carriage supporting them. The loops are then raised, moved to the next position and lowered in order to transfer the antibiotics to the diluent; the loops are rotated to mix the antibiotics with the diluent. This process is repeated as required. Inoculated broth is added to each well in the second and all subsequent rows of wells in turn, by means of a second set of 8 syringes which are mounted behind the transfer loops. This syringe assembly may be sterilised by ethylene oxide. (Multiple droppers are discussed in a little more detail on p. 63.) At the end of the process, the loops are washed in a water-bath and then manually lowered into a burner, which is automatically ignited as the loops are lowered, and held there until they are 'cherry red'.

For their studies, Goss and Cimijotti used *Staphylococcus aureus* and *Escherichia coli* in Tryptose Phosphate broth containing 50 µg/ml of triphenyl tetrazolium chloride (TTC). Their definition of minimum inhibitory concentration was that concentration of antibiotic which will inhibit the formation of a red precipitate of TTC in the bath. The method was compared with a manually performed tube dilution test. In this test the organisms were suspended in Tryptose Phosphate broth without the TTC.

The median values of eight repeat determinations of the minimum inhibitory concentration of each of 13 antibiotics for *Staph. aureus* by the standard tube dilution method and by their microtitre method were given.

(a)

(b)

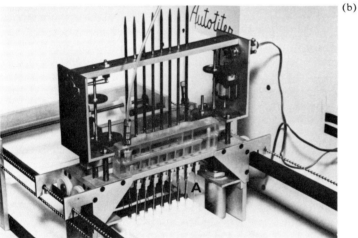

Fig. 5. The machine devised by Goss and Cimijotti for serial dilution antibiotic sensitivity testing by the microtitre method. (a) The complete apparatus. (b) Close-up view of the carriage containing the microdilutor loops. A: the syringe needles through which the inoculum is dispensed into the wells. (From Goss, W. A. and Cimijotti, E. B. (1968). Evaluation of an automatic diluting device for microbiological applications. *Appl. Microbiol.* **16,** 1414–1416.) (Photographs courtesy of Dr W. A. Goss and the American Society for Microbiology.)

Similarly, the median values for four repeat determinations of each of the 13 antibiotics for *E. coli* by both methods were given. In no case did the corresponding median values deviate by more than 1 dilution. Furthermore, they compared the median and the 25–75 percentile values for 16 determinations obtained by the Autotitre method with the median of three values obtained by the tube dilution method. As before, the results differed by no more than 1 dilution.

In view of the success of the method the authors elected to use the TTC microtitre method and demonstrated that it works equally well for *Proteus vulgaris* and *Pseudomonas aeruginosa*, although they do not present data for those two organisms. They concluded that the Autotitre apparatus is eminently suitable for routine MIC determinations; they found it very simple to use and it performed satisfactorily for 6 months without breakdown.

Gavan and Butler (1974) compared the results of tests carried out with the aid of the simple holder referred to above (p. 12) with the results obtained by a newer automatic apparatus — the Autotitre II (Ames)*—for performing (up to 8) parallel twofold serial dilutions, up to a maximum of 15.

Reproducibility of both methods was determined by repeatedly testing *E. coli* (NIHB-8) and *Staph. aureus* (NIH 289) and determining the percentage of MIC end-points which were within ±1 dilution interval. The results were as follows. With *E. coli* 197 replicates were made by the manual method and 91.4% were within ±1 dilution interval; 147 replicates were made by the semi-automatic apparatus and 97.4% were within ±1 dilution interval. With *Staph. aureus* 239 replicates were made by the manual method and 96.5% were within ±1 dilution interval; 134 replicates were made semi-automatically and 92.7% were within ±1 dilution interval. Unfortunately, the antibiotics against which these organisms were tested were not specified.

They found that with the Autotitre II equipment a sufficient number of trays can be prepared at one time to provide the routine requirements of their laboratory for at least two weeks. In consequence, experiments on storing the antibiotics at −26 deg C were carried out; the final concentration of antibiotics used in the experiments was given as 0.002 μg/ml.

A series of plates were prepared and at frequent intervals over a 14-day period, 10 plates were thawed and 5 were tested using *Staph. aureus* (ATCC 25923) and 5 using *E. coli* (ATCC 25922). They found that with drugs tested against *Staph. aureus* the median MIC end-points were the same or varied by no more than ±1 dilution interval from the overall median throughout the two-week period. The results for *E. coli* were much more variable.

Gavan and Butler felt that the semi-automatic microdilution method was 'promising'. The procedure was reproducible in 95% of the cases and thus suitable for its use in routine antibiotic sensitivity testing. They also felt they had demonstrated the practicability of preparing and storing test trays.

Although it is outside the scope of this book, one should note that Tilton et al. (1973) investigated the effects of a variety of parameters, such as inoculum concentration, methods of inoculation, media, incubation time and temperature, on the minimum inhibitory concentration test procedure. In particular, they noted that the microdilutors must be handled carefully, cleaned regularly, heated and then quenched in sterile water prior to use. They found the microdilution procedure carried out on the Autotitre II apparatus to be practical, fast and effective.

Tilton and Lieberman (1974) utilized more modern Autotitre equipment (Autotitre IV) for assaying antibiotics in body fluids. They measured the serum antibiotic levels, using *E. coli* (WHO5) as reference organism for ampicillin and cephalothin, *E. coli* (4883) as reference organism for gentamicin and *Staph. aureus* (2834) as reference organism for methicillin, by

*Miles Laboratories Inc., Ames Division, 1127 Myrtle Street, Elekhart, Indiana 46514, USA.

the manual microtitre method and also with the aid of the Autotitre IV apparatus. They performed 24 consecutive cephalothin assays and 18 consecutive ampicillin assays by both methods. They found that the assays performed on the machine gave identical results, whilst 2 of the 24 manual cephalothin assays and 3 of the 18 manual ampicillin assays gave results which were 1 dilution different from the mean dilution.

A new antibiotic susceptibility test system, the MIC2000 (Dynatech Laboratories Inc.*) is now available. It consists of a 96-channel dispenser, an inoculator and a microtitre plate viewer. Little has been published about the efficacy of the system, but Gavan (1976) has indicated that his results compare favourably with standard methods.

One could go on quoting papers on this subject but I have referred to enough to demonstrate the following points.

Very many people have used a microtitre dilution loop method for measuring susceptibility to antibiotics, and most have satisfied themselves that it is a reliable method. It is not possible to compare the work of various workers; a wide variety of different antibiotics have been used and tested against many different organisms. They have used different initial concentrations of antibiotic, different concentrations of inoculum and so on. Some have taken, or at least indicated in their published work that they have taken, much more care of the equipment than others. Some have indicated that they continually monitor the performance of the equipment, but in some cases there is not the slightest hint that this is so.

Similarly, standardization of some important parameters/processes of the procedure—for example, control organisms, media, inoculum concentration, mode of inoculation, method of incubation, length of incubation (Tilton and Newberg 1974)—has received very little attention.

Again, where there is more than 1 dilution interval difference between the microtitre method and the method by which it has been compared (usually the standard tube dilution method), no attempt has been made to decide which is the 'correct' answer. Furthermore, people seem to be quite happy to accept an agreement of the order of 90%.

It seems clear that there are many advantages in using the microtitre method; it is certainly more economic in reagents and it would appear to be no less repeatable and reliable than earlier methods. Furthermore, many of the simple aids to the microtitre method can be made to perform reasonably well, provided care is taken in their application. One can only suggest, however, that each worker evaluates the method and the equipment he himself considers adopting and satisfies himself that it is giving him sensible results, before committing himself to the substantial capital expenditure and reorganization of the laboratory which may be involved.

Multipoint inoculators

Multipoint inoculating devices are usually designed for one specific purpose, but they can almost invariably be used in two ways. The device can pick up many different inocula from an array of tubes or dishes and distribute the inocula successively to a series of tubes or test plates, or the device can pick

*Dynatech Laboratories Inc., 900 Slaters Lane, Alexandria, Virginia 22314, USA

up from one tray of inoculum and distribute it to different plates or tubes.

Lidwell (1959) developed a device for phage typing of *Staphylococcus aureus* (Fig. 6). A set of 27 wire loops (A, A^1), as designed by Tarr (1958), was mounted at each end of an arm (E) which rotated in a horizontal plane, about a vertical pillar, and stopped at 4 positions: the rod was depressed by means of a lever. In position 1, one set of loops (A) was sterilized by a 27-jet gas burner (B) and then the rod was turned to position 2 to allow the loops to cool. In position 3, they were lowered into the phage inocula (C), which were in a drilled Perspex block or other suitable vessel (in this position the second set of loops (A^1) was sterilized), and in position 4 the charged loops were lowered on to the surface of a plate previously flooded with the organism to be tested (D). Since there were two sets of loops, many plates could be inoculated very rapidly.

Fig. 6. The apparatus devised by Lidwell for phage typing. A, A': set of 27 loops. B: 27-jet gas burner. C: reservoirs of phage inocula. D: culture plate flooded with organism to be tested. The arm, E, containing the sets of loops, rotates in an anti-clockwise direction and is depressed by means of the handle, F.

A more complicated device for applying phage inoculum was devised by Zierdt et al. (1960). Twenty-six syringe barrels were fixed to a plate, and the syringe pistons were fixed to a second plate which was attached to a lead screw. An appropriate volume of each bacteriophage was dispensed, on to a previously inoculated agar plate placed under the apparatus, for each turn of the lead screw.

Another device for phage typing of staphylococci was developed by Simon and Undseth (1963). These workers used stainless steel pins mounted in a base-plate which moved up and down two vertical guide rods, and claimed that if the plate was lowered so that the pins touched the bottom of the plastic-tray-type reservoirs, '0.003 ml of phage inoculum was delivered'; they did not give evidence to support this claim.

Hill (1970) devised a phage applicator (utilizing two sets of 25 platinum wire loops) which was enclosed in an aluminium hood in which were placed two ultraviolet lamps to maintain 'cabinet sterility'. This is a precaution few workers have found to be necessary (or admitted to be necessary!) although there may well be some applications that are particularly susceptible to airborne contamination which this precaution is designed to reduce.

Various other devices, based on similar principles, have been devised. All may be used with inocula other than phage.

An aid to inoculating liquid cultures was developed by Quadling and Colwell (1964). In principle, the apparatus consisted of a brass base-plate into which 60 stainless steel needles were fixed at right angles: each needle was 9 cm long by 5 mm diameter. A guide rod was fixed in each corner of the base-plate, and the rods fitted into corresponding tubes attached to a holder containing racks of culture tubes, thus ensuring that each needle was

Fig. 7. The multiple inoculator devised by Hale and Inkley. Insert shows the simple bayonet fitting, the means by which the plate to which the stainless steel needles are permanently attached is removed from the apparatus for sterilization. (From Hale, L. J. and Inkley, G. W. (1965). A semi-automatic device for multiple inoculation of agar plates. *Lab. Pract.* **14,** 452.) (Photograph courtesy of Dr D. F. Spooner.)

properly aligned over the appropriate culture tube. A tray (or rack of tubes) containing the inocula was placed over the culture tubes and the needles were lowered, by hand, until each needle touched the inoculum. The base-plate was raised, the inoculating tray was removed and the base-plate lowered until the needles were in contact with the sterile broth in each of the culture tubes; the needles were then raised and the caps placed over the culture tubes.

Hale and Inkley (1965) developed a similar inoculating device specifically for the multiple inoculation of agar plates (Fig. 7). The inocula were placed in Oxoid aluminium test tube caps: 27 caps were placed in a jig so shaped as to keep them within the area of a 9 cm Petri dish; 2.5 mm stainless steel needles were mounted, in positions corresponding to those of the Oxoid caps, in a detachable stainless steel plate which was attached to a lever-operated press. The plate was lowered until the pins touched the inocula, it was raised, the inocula were removed and an agar plate (9 cm Petri dish) was put in position under the pins which were lowered to inoculate the agar; the detachable plate was removed for sterilization purposes.

One of the newest devices published is the 'Capillary-Action Replicator' by Mackenzie (1973), which was devised to permit the inoculation of large numbers of agar media in Petri dishes with aqueous solutions of yeasts (Fig. 8).

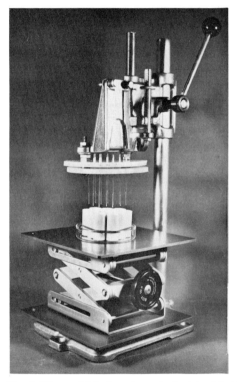

Fig. 8. The multiple inoculator devised by Mackenzie. (From Mackenzie, D. W. R. (1973). Capillary action replicator. *J. clin. Path.* **26**, 805.) (Photograph courtesy of Dr D. W. R. Mackenzie and the editor of the Journal of Clinical Pathology.)

The basis of this apparatus is a set of 9 cm long by 2 mm o.d. capillary tubes, inserted in holes in two autoclavable polypropylene sheets and held in position by means of Teflon tube collars placed over the upper end of the capillary tubes. These tubes are filled with inoculum by capillary action, and each can inoculate 25–30 agar plates per filling, although the author does not explain the mechanism by which each drop of inoculum is withdrawn from the tube—one assumes the capillary tubes are lowered to touch the surface of the agar plate.

In contrast, Fung (Fung and Hartman, 1975; Fung, 1976) used a very simple inoculator (96 stainless steel pins) to perform rapid, miniaturized microbiological tests, including gas detection in fermentation tubes, detection of motility and biochemical tests in semi-solid media and agar slants. He found the procedures required about 5% of the materials and 10% of the time required by more conventional procedures.

These are but a few examples of the many inoculators which have been produced over the years and there appears to be no limit to their application. The devices vary considerably in complexity, from 'a few nails banged into a piece of wood' (Newsom, 1975). Some are manually operated, others are semi-automatic, and the ease with which they are manipulated varies.

The volume of inoculum and accuracy with which it is transferred differ, and in any case often neither is discussed. Those factors are often not important; however, if they are important in a particular application, the appratus has to be carefully investigated from those points of view. A high degree of accuracy cannot be expected and, in any case, one would expect the accuracy and repeatability to fluctuate, particularly with age since in most devices all the inoculating loops, needles and tubes (and even the nails!) have to be frequently sterilized by heat, usually by flaming, and this process may well distort them. These components, any number of which may be fitted in any configuration (which configuration is usually chosen to suit particular inocula reservoirs or culture vessels), are usually constructed from stainless steel or platinum but other materials are not precluded, provided they are suitable from the pure bacteriological point of view and provided also they can be sterilized. Alternatively, they have to be simply replaced with freshly sterilized components.

All these factors have to be considered with each particular apparatus and application in mind. In general, however, the advantages of using a multipoint inoculator when one has to distribute an inoculum or many different inocula successively to a series of tubes or plates are incontrovertible: the techniques are significantly faster and less costly than other methods, even when quite small numbers of tests have to be carried out, and variations in the volume of inoculum transferred are reduced to a minimum.

References

Pipetting small volumes of fluids

Dropping pipettes

Donald, R. (1913). A method of counting bacteria in water. *Lancet* **1**, 1447.

Donald, R. (1915). A method of drop-measuring liquids and suspensions. *Lancet* **2**, 1243.

Fitzgerald, S. C., Fuccillo, D. A., Moder, F. and Sever, J. L. (1974). Utilization of a further miniaturized serological microtechnique. *Appl. Microbiol.* **27**, 440.

Gavan, T. L. and Town, M. A. (1970). A microdilution method for antibiotic susceptibility testing. *Amer. J. clin. Path.* **53**, 880.

MacLowry, J. D., Jaqua, M. J. and Selepak, S. T. (1970). Detailed methodology and implementation of a semi-automated serial dilution microtechnique for antimicrobial susceptibility testing. *Appl. Microbiol.* **20**, 46.

Miles, A. A. and Misra, S. S. (1938). The estimation of the bactericidal power of the blood. *J. Hyg. (Lond.)* **38**, 733.

Tilton, R. C., Lieberman, L. and Gerlach, E. H. (1973). Microdilution antibiotic susceptibility test: examination of certain variables. *Appl. Microbiol.* **26**, 658.

Wilson, G. S. (1922). The proportion of viable bacteria in young cultures with especial reference to the technique employed in counting. *J. Bact.* **7**, 405.

Wilson, G. S. (1935). The bacteriological grading of milk. *Spec. Rep. Ser. med. Res. Coun. (Lond.)* No. 206.

Semi-automatic pipettes

Bousfield, I. J., Smith, G. L. and Trueman, R. W. (1973). The use of semi-automatic pipettes in the viable counting of bacteria. *J. appl. Bact.* **36**, 297.

Ellis, K. J. (1973). Errors inherent in the use of piston activated pipettes. *Analyt. Biochem.* **55**, 609.

Joyce, D. N. and Tyler, J. P. P. (1973). Accuracy, precision and temperature dependence of disposable tip pipettes. *Med. Lab. Technol.* **30**, 331.

Robinson, S. M. and Johnson, K. R. (1974). The assessment of the accuracy and precision of semi-automatic pipettes. *Med. Lab. Technol.* **31**, 213.

Transferring small volumes of fluids

Diluting loops

Ashcroft, J., Platt, G. S. and Maidment, B. J. (1971). The accuracy of the microtitre technique. *Med. Lab. Technol.* **28**, 129.

Harwick, H. J., Weiss, P. and Fekety Jr., F. Robert (1968). Application of microtitration techniques to bacteriostatic and bactericidal antibiotic susceptibility testing. *J. Lab. clin. Med.* **72**, 511.

Marymont, J. H. and Wentz, R. M. (1966). Serial dilution antibiotic sensitivity testing with the microtitrator system. *Amer. J. clin. Path.* **45**, 548.

Sever, J. L. (1962). Application of a microtechnique to viral serological investigations. *J. Immunol.* **88**, 320.

Takatsy, Gy. (1955). The use of spiral loops in serological and viral micro-methods. *Acta microbiol. Acad. Sci. Hung.* **3**, 191.

Semi-automatic diluting apparatus

Chitwood, L. A. (1969). Tube dilution antimicrobial susceptibility testing: efficacy of a microtechnique applicable to diagnostic laboratories. *Appl. Microbiol.* **17**, 707.

Gavan, T. L. and Town, M. A. (1970). A microdilution method for antibiotic susceptibility testing. *Amer. J. clin. Path.* **53**, 880.

MacLowry, J. D. and Marsh, H. H. (1968). Semiautomatic microtechnique for serial dilution–antibiotic sensitivity testing in the clinical laboratory. *J. Lab. clin. Med.* **72**, 685.

MacLowry, J. D., Jaqua, M. J. and Selepak, S. T. (1970). Detailed methodology and implementation of a semi-automated serial dilution microtechnique for antimicrobial susceptibility testing. *Appl. Microbiol.* **20**, 46.

More sophisticated diluting apparatus
Gavan, T. L. (1976). Quantitative microdilution susceptibility tests using an automated test system — MIC-2000. In: *Proceedings 2nd International Symposium on Rapid Methods and Automation in Microbiology.* Eds. H. H. Johnston and S. W. B. Newsom. Learned Information (Europe) Ltd., Oxford.
Gavan, T. L. and Butler, D. A. (1974). An automated microdilution method for antimicrobial susceptibility testing. In: *Current Techniques for Antibiotic Susceptibility Testing.* Ed. A. Balows. Charles C. Thomas, Springfield, Illinois.
Goss, W. A. and Cimijotti, E. B. (1968). Evaluation of an automatic diluting device for microbiological applications. *Appl. Microbiol.* **16**, 1414.
Tilton, R. C. and Lieberman, L. (1974). Microdilution assay of antibiotics in body fluids. *Ann. clin. Lab. Sci.* **4**, 178.
Tilton, R. C. and Newberg, L. (1974). Standardization of the microdilution susceptibility test. In: *Current Techniques for Antibiotic Susceptibility Testing.* Ed. A. Balows. Charles C. Thomas, Springfield, Illinois.
Tilton, R. C., Lieberman, L. and Gerlach, E. H. (1973). Microdilution antibiotic susceptibility test: examination of certain variables. *Appl. Microbiol.* **26**, 658.
Multipoint Inoculators
Fung, D. Y. C. and Hartman, P. A. (1975). Miniaturized microbiological techniques for rapid characterization of bacteria. In: *New Approaches to the Identification of Microorganisms.* Eds. C-G. Hedén and T. Illeni. John Wiley, New York and Chichester.
Fung, D. Y. C. (1976). Miniaturized microbiological techniques. In: *Proceedings 2nd International Symposium on Rapid Methods and Automation in Microbiology.* Eds. H. H. Johnston and S. W. B. Newsom. Learned Information (Europe) Ltd., Oxford.
Hale, L. J. and Inkley, G. W. (1965). A semi-automatic device for multiple inoculation of agar plates. *Lab. Pract.* **14**, 452.
Hill, I. R. (1970). Multiple inoculation technique for rapid identification of bacteria. In: *Automation, Mechanization and Data Handling in Microbiology.* Eds. A. Baillie and R. J. Gilbert. Academic Press, New York and London.
Lidwell, O. M. (1959). Apparatus for phage typing of *Staphylococcus aureus. Bull. Hyg. (Lond.)* **34**, 818.
Mackenzie, D. W. R. (1973). Capillary-action replicator. *J. clin. Path.* **26**, 805.
Newsom, S. W. B. (1975). Easy, economic, typing of enterobacteria. In: *New Approaches to the Identification of Microorganisms.* Eds. C-G. Hedén and T. Illeni. John Wiley, New York and Chichester.
Quadling, C. and Colwell, R. R. (1964). Apparatus for simultaneous inoculation of a set of culture tubes. *Canad. J. Microbiol.* **10**, 87.
Simon, H. J. and Undseth, S. (1963). Simple method for phage typing of staphylococci. *J. Bact.* **85**, 1447.
Tarr, H. A. (1958). Mechanical aids for the phage typing of *Staphylococcus aureus. Bull. Hyg. (Lond.)* **33**, 698.
Zierdt, C. H., Fox, F. A. and Norris, G. F. (1960). A multiple-syringe bacteriophage applicator. *Amer. J. clin. Path.* **33**, 233.

2
Devices for the enumeration of bacteria

Devices for the enumeration of bacteria may be placed in one of two main categories: those for counting colonies of bacteria on or in solid media, and those for measuring concentrations of bacteria in suspension.

Colony counters

For counting colonies in capillary tubes

A delightfully simple technique was devised by Yanagita (1956). In this technique the bacterial suspension, of concentration 10^3-10^4 organisms/ml, is mixed in melted nutrient agar (0.5 ml of suspension to 4.5 ml of agar) and a quantity of the mixture is sucked into each of 3 capillaries of internal diameters between 1.3 and 1.6 mm. There are two marks on each capillary, the distance between them being between 250 and 380 mm, depending on the internal diameter of the capillary, so that the volume of agar between these two marks can be calculated. The capillary has a thick wall, so that when the capillary is passed between a very simple lamp/lens system (Fig. 9) the colonies which grow during incubation, usually of 24 h duration, are suitably

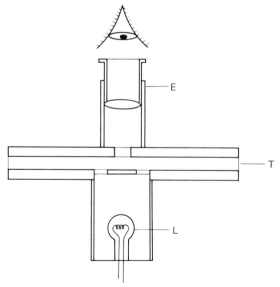

Fig. 9. Sketch showing principle of the apparatus devised by Yanagita for counting colonies in capillary tubes. E: eye-piece. L: lamp. T: metal tube in which capillary tube containing the culture is placed.

focused; as the tube is moved through the device the number of colonies between the lines inscribed on the capillary are counted. Viable counts obtained by this method were reported as being practically identical with those obtained by plate counts, although the evidence was not given. Yanagita found that it was very easy to count the colonies, that he could count up to 10^3 cells (colonies?)/ml, and that the method was very accurate. However, it has limitations, since it can only be used with anaerobes and facultative anaerobes that do not form gas. (In early experiments he attempted to use it, for example, with *Escherichia coli*, *Staphylococcus aureus* and *Lactobacillus casei*.)

This was one of the earliest of a large number of aids to methods for counting viable bacteria that may be described as capillary tube methods, reviewed in some detail by Hartman (1968), which were the forerunners of the system devised by Bowman et al. (1967). This is a rather more sensitive method for counting viable bacteria (and also for determining antibiotic sensitivities) than classical photometric methods, which we discuss later.

Specially prepared agar is melted and the specimen is added to the agar and mixed thoroughly. The end of a capillary tube is dipped into the agar and the tube is filled by capillary action; both ends of the tube are sealed. Approximately 150 µl of agar is needed to fill the tube, which is left until the agar has solidified.

The capillary tube is scanned by a linear tungsten filament lamp, which is focused onto the centre of the capillary (Fig. 10). An objective collects the

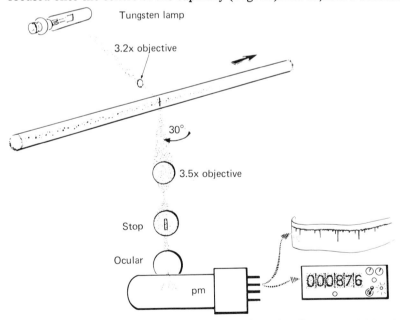

Fig. 10. Sketch showing principle of the apparatus devised by Bowman and his colleagues for counting colonies in capillary tubes. (From Bowman, R. L. et al. (1967). Capillary-tube scanner for mechanized microbiology. Science, **158**, 78–83) (Copyright 1967 by the American Association for the Advancement of Science. Drawing courtesy of Dr Philip Blume.)

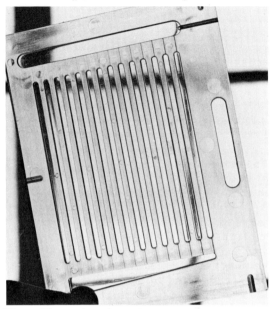

Fig. 11. The 14–channel, moulded plastic plate for use in the apparatus devised by Blume and his colleagues. (*From* Blume, P. et al. (1975). Automated antibiotic susceptibility testing. In: *Automation in Microbiology and Immunology.* Eds. C-G. Hedén and T. Illeni. John Wiley.) (Photograph courtesy of Dr Philip Blume and the Publisher.)

light scattered within the capillary and the light is passed through a stop to a photomultiplier tube. The image of the source is 50 nm wide. The instrument counts, electronically, light pulses scattered from growing colonies and from other points, such as the joint between the agar and the capillary. The tube is scanned at intervals and growing organisms produce new or larger pulses than those obtained on a previous scan, but other scattering points produce constant pulses. The background scatter is such that it obscures small signals and, therefore, the organism has to divide several times before it can be detected. The threshold level was set at 4 times the background and the minimum number of organisms that forms a recognizable microcolony is of the order of 20.

At high concentrations, the true count is higher than the measured count because two or more colonies in the area formed by the image of the source are counted as one. The effect becomes significant in the case of *Escherichia coli* at a concentration of 10^5/ml. In addition, at high concentrations the supply of nutrients is exhausted more quickly than if fewer organisms are present. Growth curves for *E. coli, Streptococcus, Proteus* and *Staphylococcus* were obtained, and in the case of *E. coli* a quantitative estimate of the number of viable cells present in a capillary was made within 5 h. A rough estimate of the number of viable cells present was made in 2 h. Longer periods were required for the other organisms; much slower growing organisms require overnight incubation.

To determine antibiotic sensitivities, the antibiotic was mixed in the agar during its preparation and before the culture was added. Growth curves for

Klebsiella in different concentrations of sodium colistimethate were obtained, and the difference in growth rates of specimens with large differences in concentration of antibiotic was easily measured, but the resolution was poor.

Bowman and his co-workers found that the preparation and handling of the capillary tubes was very laborious and possible modifications to the technique were under investigation. This work had considerable potential; not only is it a very sensitive method for detecting growth of bacteria in suspension, it might have led to means of producing some bacteriological test results on the same day as the specimen is sent to the laboratory.

Unfortunately, however, although a great deal more work has been devoted to this device — particularly to devising a moulded plastic plate, with 14 parallel rectangular channels 2.2 mm wide × 1.8 mm deep (Fig. 11) which when covered with adhesive film constitute the equivalent of 14 capillary tubes (Blume et al., 1975), as a means of eliminating the handling of large numbers of capillary tubes — and although the apparatus is now commercially available (American Instrument Company⋆), it does not seem to have become as widely used as one might have expected.

A very similar device to that of Bowman and his colleagues was devised and evaluated by Schoon et al. (1970). They used 100 μl disposable pipettes, filled, 10 at a time, by placing them in a holder and applying a negative pressure to one end when the other was dipped into the molten agar in which the organisms had been mixed. The holder is, of course, sterilizable. When the agar has solidified, the capillaries are removed from the holder and placed into a rack. Racks hold up to 21 tubes which may, therefore, be loaded into the optical system together. The authors stated that 10 different samples with 10 tubes each can be prepared by one operator in less than 45 min, including the time taken to prepare the culture.

The optical scanning system is very similar to that used by Bowman and his colleagues, but, whereas they attempted to detect colonies with diameters of the order of 5 μm, Schoon and his colleagues were concerned only with colony diameters in the range 50–100 μm, because an early estimate of colony numbers was not required in their study of the dynamics of microbial populations. The capillary is scanned in both directions and the total count is displayed on counting tubes or, of course, is printed out.

The instrument was evaluated in two ways. First by plotting the results obtained (from an average of 20 tubes) by the instrument with those obtained by a direct count carried out under the microscope. The authors said that 'a reasonably good response' was obtained with *Escherichia coli*, *Lactobacillus casei* and *Saccharomyces cerevisiae* if the colony count ranged from 10 to 400, but there were greater errors at the lower end of that range than elsewhere. The second method was to compare the viable count of *E. coli* obtained by the pour plate, the spread plate and the scanner methods. Twenty specimens, each from the same culture, were made up for each method, and all three methods gave essentially the same results; $1.5 \pm 0.11 \times 10^9$/ml, $1.48 \pm 0.55 \times 10^9$/ml and $1.46 \pm 0.046 \times 10^9$/ml for the pour plate method, the spread plate method and the scanning method respectively. These figures were

⋆American Instrument Company, 8030 Georgia Avenue, Silver Spring, Maryland 20910, USA.

within the 95% confidence limit.

However, this scanning method is not very satisfactory. A variety of problems arise. Great care is necessary in preparing the tubes in order to ensure that gas bubbles are not created. Excessive agitation creates gas bubbles which the machine cannot distinguish from colonies. Tubes that have not been washed properly can also give misleading results and, of course, all the problems due to coincidence and very high concentrations of organisms discussed above remain. Furthermore, the result is related to the number of colonies, their diameters, and also to their opacity. Schoon devotes a considerable amount of space to discussing these errors. The problem is, of course, that although it is relatively easy to determine where errors arise, it is not so easy to eliminate them.

As we shall see, this applies to all similar scanning techniques for counting colonies. It appears that one can produce a method that is accurate to something of the order of 85%, but pushing the accuracy beyond that is very difficult, and it is likely to be prohibitively expensive.

Despite the degree of sophistication of the Bowman and the Schoon methods, the simple capillary tube methods are still being devised. For example, Nakamura et al. (1974) used one method for counting viable cells of *Bifidobacterium bifidum* grown in a solid medium. They devised a little agar puncher for sampling a constant volume of the medium containing the cells; this is an aluminium cap with a hole punched in the bottom through which a glass tube is placed. A rubber bulb is attached to the other end of the glass tube so that when the cap has punched a little disc of agar it can be sucked up with the aid of the rubber cap and discharged into a homogenizer. The specimen contains 0.7 ml of agar and this is added to 9.3 ml of sterile physiological saline. A range of dilutions of the homogenate were made and the number of cells in each counted in a capillary tube. It would appear that these very simple aids will retain their usefulness for some years to come.

For counting surface colonies

A very simple device for counting colonies on agar plates is a manually held probe which is lightly touched on each colony in turn. A second probe, which is connected to the first probe through an impulse counter and a power supply, and which is often a simple crocodile clip, is connected to the agar so that each time a colony is touched by the first probe, the circuit is closed (Fig. 12) and the impulse counter registers a count. A whole host of such devices are in use together with a range of agar plate illuminators (see, for example, Moore and Taylor, 1950) which are used to assist the operator in this process.

One normally associates this technique with counting colonies obtained by distributing a liquid culture over the whole surface of the agar plate, but it can in fact be used for counting colonies in the drop technique (Miles and Misra, 1938) which has hitherto been very widely used; the method — in which small drops of samples, usually of the order of 20 µl dispensed with a calibrated dropping pipette, from at least ten different dilutions of the culture are put on to the surface of an agar plate and the colonies occurring after incubation are counted — is well known.

One disadvantage with the probe device described above is that the colonies

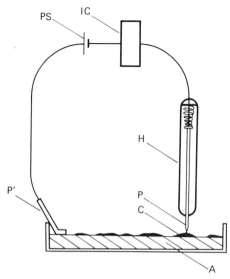

Fig. 12. Sketch illustrating a simple surface colony counting device. A: agar. C: surface colony. IC: impulse counter. PS: power supply. P: probe, in handle, H, touching a colony. P': second probe.

are touched by the probe and it is necessary, therefore, for it to be sterilizable. There are some applications in which this method is unsuitable or, at least, undesirable.

An alternative is to use a similar device but one which makes a mark on the bottom of the plate at a point immediately over the colony and at the same time actuates the impulse counter. Goebel and Blum (1950) devised such a device. This is a spring-loaded holder for a lead pencil, which is held in such a way that when its tip touches the plate and pressure is applied, a simple switch built into the holder is actuated to operate the counter. Many devices of this nature are also very widely used and are readily available commercially.

The above methods are, however, very time consuming and various ways to speed up the counting process have been suggested. A comparative method in which the specimen with an unknown colony count is compared with standard specimens has been devised by Murphy and Tucker (1970). Since this can hardly be described as a technological aid, it is outside the scope of this book, but I mention it here because Murphy and Tucker claim that it is possible to rapidly assess the number of colonies on plates with up to 1,000 colonies with an error of ± 10%, and that the speed of the process is independent of the number of colonies present.

Very sophisticated electronic counting techniques have been used for automatically counting colonies on solid agar plates. Such an apparatus was described by Alexander and Glick (1958). This was an electronic scanning device in which a cathode ray tube flying spot was focused onto a culture and variations of light transmitted during the scan were measured on a photomultiplier. A dual light/photomultiplier anti-coincidence network was used.

Despite this, the chief source of error was due to coincidence. Bouffant and Soule (1954) discussed such errors in detail. Briefly, they fall into four categories: fractional interception errors, when one particle is intercepted by two or more passages of the scanning beam; overlap errors, when particles overlap in the projected image of the scanning beam and are counted as one; coincidence errors, when two particles are so close together that the apparatus cannot discriminate between them; and sensitivity errors, when the apparatus cannot distinguish a particle from the background.

Alexander and Glick's apparatus was evaluated by Malligo in 1965; the time between publication of these two papers suggests many difficulties were encountered. Malligo used *Bacillus subtilis* and *Serratia marcescens* plated on peptone agar and tryptone agar respectively. The agar had to be transparent and therefore blood agar could not be used. Plates were prepared so that the number of colonies ranged between 30 and 300, and the colonies were kept away from the edge of the plate. They counted 3,000 plates for each specimen on the machine and manually. In general, the machine gave a lower count than the manual method if more than 100 colonies were present and a higher count than the manual method if less than 100 colonies were present. These discrepancies were attributed to colonies being on the periphery of the scanning area, touching and overlapping colonies and optical imperfections in agar plates. Constant relationships between machine and visual counts were sought and an equation was derived for each species that related the machine count to the visual count within the 95% confidence limit.

The difficulties that must have been encountered during this work were not discussed, but it appeared (in 1958) that within the obvious limitations of the method one would be able to improve the technique so that one would eventually be able to count colonies on an agar plate in this way.

In consequence, since that time a lot of work on automatic colony counters has been carried out, in particular by commercial organizations. There are now some very sophisticated electronic devices, costing several thousand pounds or even tens of thousands of pounds, commercially available.

Goss et al. (1974) published an evaluation of one such electronic high speed scanning colony counter (Artek Systems Corporation*). *Staphylococcus epidermidis* and *Escherichia coli* from skin samples, which were plated on Trypticase glucose extract agar (with added lecithin and polysorbate 80), were used to prepare surface colonies on Trypticase Soy agar. Pour plates were also produced. All plates were incubated at 37 deg C for approximately 48 h and manual plate counts were performed. The specimens were also counted on the automated colony counter.

In some initial experiments they discovered that the device could detect both surface and subsurface colonies and that a colony diameter of 0.3 mm was the minimum practical size of colony that the instrument could reliably resolve; they also found that 'occasional' artefacts in the agar produced erroneously high counts so that plates of questionable quality had to be rejected.

*Artek Systems Corporation, Farmingdale, New York, USA.

A variety of experiments were performed. The counter reproducibility was found to be related to the number of colonies in the plates. Repetitive counts (between 10 and 20) on plates the position of which was unchanged varied by 3%, 2% and 1% for colony densities 30–100, 200–300 and 600–1,000 respectively. If the plates were moved between each count, variations were 12%, 4% and 3% respectively.

In comparing the results obtained by the automatic colony counter with the results obtained by the manual technique, it was found that there was a significant discrepancy, particularly when the colony count was greater than 100. However, when the data were plotted on a log–log scale, there was an almost linear relationship over the range of 10–1,000 colonies, although comparatively few plates with colonies greater than 300 were examined; in those cases the manual count was an estimate based on counts of representative subareas. They considered a colony count of up to 400 to be a practical range and that the accuracy of counting colonies was of the order of 90–95%.

Yourassowsky and Schoutens (1974, 1975) used a commercially available colony counter (Zeiss Micro-Videomat*) for counting and determining the size of colonies developing on culture plates with inhibition zones produced by antibiotic discs of penicillin G and sulphadiazine, as well as for counting colonies distributed over the whole surface of a plate. This is a very complicated and, in principle, similar piece of equipment to that used by Goss and his colleagues, and I do not intend to discuss this work in detail.

It is interesting to note, however, that Yourassowsky and Schoutens stated that differences existed between automated and manual counts (in the case of *Klebsiella pneumoniae* the difference was 20%) and the differences were related to touching and overlapping colonies and to the fact that the machine cannot resolve clusters of colonies. They suggested, however, that increased precision of counting by the machine may be obtained by growing the cultures in triple layers, by which they mean two layers of agar with a third layer in which the culture has been mixed sandwiched in between; the total thickness of all three layers should not exceed 6 mm so that colony numbers of 10^3–10^4 per plate may be accurately counted. Even so, they suggested that correlation with the Kirby–Bauer procedure is not good and requires further investigation.

Although there are several similar devices commercially available, very few papers like those of Goss et al. and Yourassowsky and Schoutens describing evaluations of these machines have been published, and on the available evidence it would appear that these devices have a long way to go before they can be regarded as sufficiently reliable for most applications. (It seems to me significant that Goss and his colleagues recommend that their particular system, although it has potential, should be evaluated for each specific application potential users propose.) Clearly, extreme caution is advisable before spending a substantial sum of money on such a device.

Despite the encouraging way in which some people talk about electronic automatic surface (and subsurface) colony counters, I remain sceptical about their performance and also about their use in microbiology. I have yet to see a

*Carl Zeiss, 7082 Oberkochen, Postfach 35/36, West Germany.

convincing demonstration of one and cannot help but wonder if much progress has been made since Alexander and Glick produced their machine in 1958. Certainly electronic techniques have been developed to such an extent that the signal obtained from the scanner can be displayed in a multitude of very fancy ways. But the real test is, is the colony count really accurate? How can it be if one gets a different answer if one turns the Petri dish around and repeats the count?

We have already seen that although designed for surface colony counting, or at least first used in bacteriology for that purpose, some of these devices have been used for counting subsurface colonies, as obtained in pour plate and similar methods. This applies particularly to the simple probe type devices; the Goebel and Blum probe is suitable for such applications.

Other colony counting techniques

A technique in which a probe has been used for counting colonies is a miniature pour plate technique, which was described by Sharpe and Kilsby (1971) and Sharpe (1973). In principle, the technique involves producing 0.1 ml droplets of molten agar in which the specimen has been suspended.

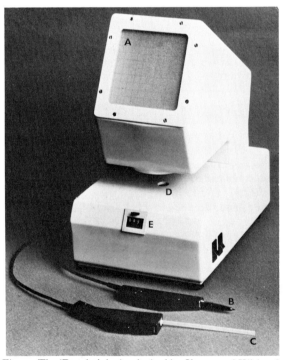

Fig. 13. The 'Droplet' device devised by Sharpe and Kilsby. A: screen on to which image of each agar drop is projected. B: probe for counting colonies (by touching screen over the image of each colony). C: probe through which molten agar is sucked up/ejected. D: orifice over which the agar drop is placed for colony counting. E: counter. (Photograph courtesy of Seward Laboratory.)

The special diluting/dispensing device contains two pumps operated by separate pedals; unfortunately, details of the pumps were not given. One pump operates to suck up and eject 1.5 ml of fluid and the other operates to eject 0.1 ml of fluid through a sterile Pasteur pipette, or even a drinking straw.

Clearly there are a variety of ways in which this device may be used but one method is to suck up 1.5 ml of the specimen suspended in molten agar and to dispense five 0.1 ml droplets in a line on a Petri dish; the remaining 1 ml of suspension may be discharged into a vessel containing 9 ml of diluent agar to give a 1 in 10 dilution, or 0.1 ml of the suspension may be discharged into 9 ml of diluent, to give a 1 in 91 dilution, so that a series of drops of different dilutions of suspension may be dispensed: Sharpe claimed that the difference between the 1 in 91 dilution and the 1 in 100 dilution is unlikely to be noticed but, clearly, the volumes of diluent may be adjusted to eliminate this error. The Petri dish containing the rows of droplets may be placed on a simple viewer (Fig. 13) so that the droplet under examination is placed over the illumination hole. The droplet acts as its own condenser, so that no focusing is required, and it is magnified by a factor of 10 so that it appears on the viewing screen to be about the size of the 90 mm Petri dish. The colony images may be touched with a felt tipped pen or probe for counting, as discussed earlier.

This appears to be a very useful device and is commercially available (Seward Laboratory*). However, problems have emerged. The heat generated by the lamp in the viewer, which is quite close to the Petri dish, is such that there is a tendency for the agar droplet to melt. There is also a tendency for the plastic Petri dish to become distorted. Furthermore, the accuracy of dispensing the droplets has been found to be erratic in some commercial models of the machine.

This technique was designed specifically for performing bacterial counts on food samples, and in that application, in which the suspensions were prepared by homogenizing 0.01 kg of sample in 90 ml of 0.1 % peptone solution for 2 min, it was found to compare favourably with the classical pour plate technique.

Sharpe also found that the visibility of the colonies in the droplets at 24 h is rather better than the visibility of colonies in pour plates at 48 h. He also found that food debris was easily distinguishable from bacterial colonies in the droplets. Furthermore, it was found that a technician can process approximately three times as many samples per day by this technique as may be processed by the pour plate technique. Also, the shorter incubation time is advantageous. Clearly, there are applications of this technique in many branches of bacteriology other than food bacteriology.

Another machine designed for the food bacteriology field, but which has application in other branches of bacteriology, was described by Gilchrist et al. (1973). In this machine an undiluted liquid culture is slowly discharged through a thin tube, which is connected to a syringe in which the culture is contained, as the carriage to which the syringe assembly is attached moves along a radius of a rotating Petri dish agar plate. Thus the culture is

*Seward Laboratory, UAC House, Blackfriars Road, London SE1 9UG, UK.

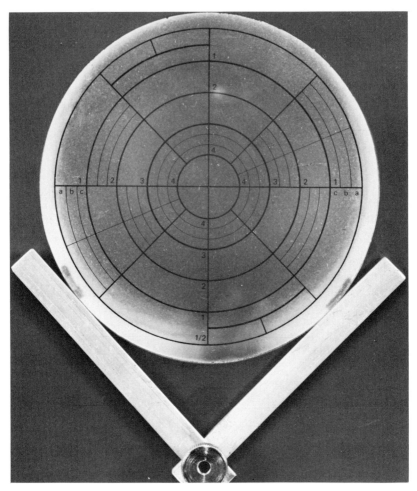

Fig. 14. The viewing box, with graticule delineating the sections into which the culture plate is divided for counting surface colonies by the method devised by Campbell and Gilchrist. (Photograph courtesy of Dr J. E. Campbell.)

discharged onto a spiral curve on the agar: the tip of the tube remains in contact with the agar throughout.

The volume of culture (the concentration of which is restricted to the range $600\text{–}10^7$ organisms/ml) discharged is controlled and decreases with time, so that different volumes, and thus different numbers of organisms, are discharged onto different areas of the plate: a total of approximately 30 μl is used. Thus, on incubation there is one area of the plate, divided into sections as shown in Fig. 14, on which the optimum number (approximately) of colonies for counting is found. One counts the colonies in that section and, since the volume of culture discharged onto each section is known, one can calculate the concentration of organisms in the original sample: the colonies may be counted, with the aid of a viewing box, in the conventional way.

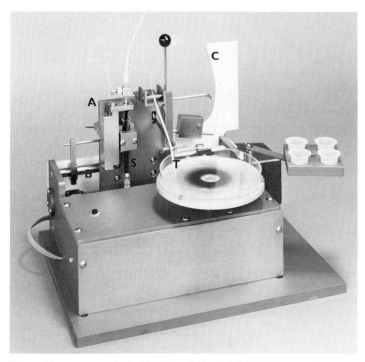

Fig. 15. The apparatus devised by Campbell and Gilchrist for preparing culture plates for surface colony counting. A: carriage containing culture-dispensing mechanism. B: rod on which carriage moves. C: cam the shape of which determines the rate at which culture is discharged. S: culture-dispensing syringe. T: culture is dispensed from this tube as it moves along a radius of the rotating dish. (Photograph courtesy of Dr J. E. Campbell.)

(The plate is rotated and the carriage is moved by means of simple mechanical mechanisms (Fig. 15). The syringe barrel (S) is fixed to the carriage (A). As it moves across the plate the syringe plunger is depressed (through a rack and pinion) by one end of a lever the other end of which traces a curve defined by the shape of a cam (C), with which the lever remains in contact. Thus, the shape of the cam determines the rate, and the variation of the rate, at which the plunger is depressed, and thus the rate at which the sample is discharged.)

This device, which is used for studying, in particular, organisms found in milk and mashed potatoes, was tested using *Bacillus subtilis, Pseudomonas aeruginosa, Staphylococcus aureus* and *Lactobacillus casei.* The results obtained with the machine were compared with those using the pour plate method. Unfortunately, there were some wide discrepancies in the percentage difference of the geometric means of 18 repeat tests by both methods; they ranged from 0.5% (for *E. coli* and *Staph. aureus*) to 36.6% (for one *E. coli* test). The performance of a more recent version of the machine (Campbell and Gilchrist, 1973) is rather better.

I like this idea; it may well have advantages over the pour plate method, and clearly it has applications in medicine.

(a)

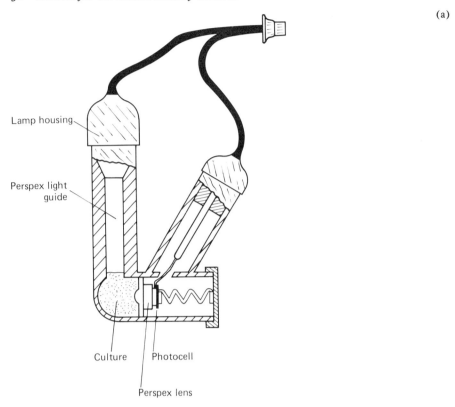

Lamp housing

Perspex light
guide

Culture

Photocell

Perspex lens

(b)

Fig. 16. The lamp/photocell system used in the nephelometer devised by Cobb and his colleagues. (a) Sketch showing principle of the totally enclosed cell. (b) Photograph showing 12 independent cells in a water-bath. (Sketch and photograph courtesy of Dr D. F. Spooner.)

Read et al. (1974) and Briner et al. (1975) briefly referred to an electronic scanner for automatically counting the colonies on the plates produced by this machine, but little detail was given. It is still under development (Campbell, 1976).

Sharpe (1973) devised a machine for automatically preparing pour plates of a range of dilutions of a culture. This is a very complex device and is outside the scope of this book. It is, nevertheless, an ingenious machine which the mechanically minded may well wish to look at more closely by reading the literature.

Measuring concentrations of bacteria in suspension

Simple photometric devices

Photometric devices (Snell and Snell 1948) have been widely used in experimental bacteriology for many years — at least since 1933, when Alper and Sterne measured growth curves of *Salmonella gallinarum* and Pulvertaft and Lemon measured opacity curves of *Escherichia coli* growing in beef broth and in Lemco broth. They did not define opacity. Brief reviews of the applications of photometric methods to bacteriology are given by Norris (1959), Kavanagh (1963), Meynell and Meynell (1965) and Robrish et al. (1971). Many devices, such as the Vitatron universal photometer-densitometer (Vitatron Scientific Instruments★) and the Unicam spectrophotometer (Pye Unicam Limited†), are commercially available and new instruments are still being devised.

Cobb et al. (1970) described a nephelometer in which 12 independent glass cells containing lamp, photocell and culture (Fig. 16) are rocked in a specially constructed water-bath and continuously monitored for turbidity. A growth curve for each of the 12 cultures is printed out on a single chart.

Bergan (1975) devised a similar device for the simultaneous monitoring of 12 cultures, but whereas Cobb's device was specially constructed, Bergan used a standard spectrophotometer (Zeiss PMQ11§). In this system the standard cell-holder is replaced by a movable aluminium slide, which moves on wheels running on rails, on which the 12 separate cells containing the cultures are placed in a row. Each cell is moved automatically, in turn, to the measuring position of the spectrophotometer, by means of an electric motor and microswitches located in line with each cell (Fig. 17).

Glass cylinders 8 × 2 cm are mounted on top of commercial 2 × 2 cm rectangular cells and the complete units are placed in heated holders. The cultures are stirred by means of bubbling air through them, and also by means of magnetic stirrers placed underneath the cells.

Tennant and Withey (1970) modified two colorimeters, the Bausch & Lomb‖ Spectronic 20 regulated model and the EEL¶ Spectra, to produce

★Vitatron Scientific BV, Kanaalweg 24, PO Box 76, NL 6210 Dieren, The Netherlands.
†Pye Unicam Ltd., York Street, Cambridge CB1 2PX, UK.
§Carl Zeiss, 7082 Oberkochen, Postfach 35/36, West Germany.
‖Bausch & Lomb Incorporated Optic Centre, 1400 North Goodman Street, Rochester, New York, 14602, USA.
¶Corning Medical, Corning Ltd., St Andrews Works, Halstead, Essex, UK.

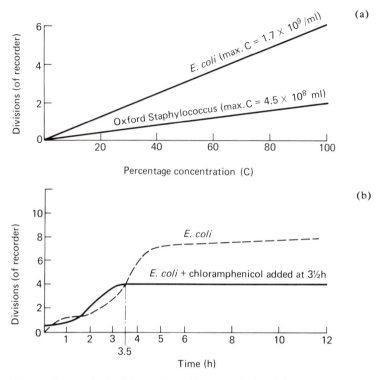

Fig. 20. Curves obtained from the turbidimeter designed by Watson and his colleagues. (a) Calibration curves for *E. coli* and the Oxford *Staphylococcus*. (b) Growth curves of *E. coli*. (From Watson, B. W. et al. (1969). A simple turbidity cell for continuously monitoring the growth of bacteria. *Phys. in Med. Biol.* **14**, 555–558.) (Copyright The Institute of Physics.)

cuvette is filled with fresh broth. Subsequently it is filled with a standard culture, which is a fully grown culture stabilized with formalin, and the recorder is adjusted to read 100 divisions. The instrument then responds linearly with cell concentration for the particular test tube and the particular organism. This process has to be repeated for each tube and for each organism used. The authors suggested that if a special tube could be manufactured, it may not be necessary to repeat this procedure for every tube.

The equipment was calibrated using *Escherichia coli* and the Oxford *Staphylococcus*, the maximum concentrations of which were 1.7×10^9 and 4.5×10^8 organisms/ml respectively; the curves obtained are shown in Fig. 20a. Growth curves of *E. coli* with and without an inhibitory concentration of chloramphenicol are shown in Fig. 20b.

A further development of this device was described by Mackintosh et al. (1973). This is a multichannel system devised for performing antibiotic sensitivities and antibiotic assays. Two problems found with the simple apparatus, namely temperature fluctuations and bacterial growth on the walls of the test tube, have been eliminated in this device, which has 12

PHOTOCELL   CULTURE TUBE

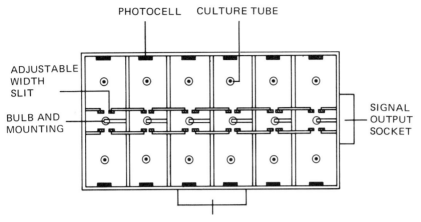

ADJUSTABLE
WIDTH
SLIT

BULB AND
MOUNTING

SIGNAL
OUTPUT
SOCKET

THERMOSTAT ELEMENT

Fig. 21. The layout of the lamps and the photocells in the multichannel turbidimeter designed by Mackintosh and his colleagues. (Drawing courtesy of Prof F. O'Grady.)

independent channels. Twelve of the simple turbidity cells have been arranged inside a single temperature-controlled box. There are 6 light bulbs, connected (electrically) in parallel, each of which illuminates two slits, the slit and photocell being separated by the same distance as in the single cell machine (Fig. 21). Consequently, temperature equilibrium is necessary before measurements are made.

Norris et al. (1970) devised an automatic growth recorder for recording bacterial growth curves. Samples are withdrawn from the culture by means of a 5 ml gas-tight syringe. The syringe is housed in a light-tight box which fits between the light source and the photomultiplier of a Vitatron* spectrophotometer. The syringe is arranged so that the light passes through that part of the barrel which is filled with the liquid when the plunger is withdrawn. Samples are returned to the culture vessel after each measurement. The movement of the syringe plunger on the inside of the barrel prevents wall growth, a problem frequently encountered in devices of this sort.

This is a closed system which eliminates many of the problems encountered when samples have to be taken over long periods. The sample rate can be varied and the optical density of the culture over a period of many days may be easily obtained. The spectrophotometer is otherwise unmodified.

A device devised to continuously record the turbidity of a continuously growing culture, but which clearly could be used for a variety of other purposes, was devised by Robrish et al. (1971). This is a fibre optic probe which is used in conjunction with a standard spectrophotometer.

In place of the normal cuvette assembly, one uses an assembly containing a mirror which deflects the incident light to the end of a bundle of fibre optics. The light is transmitted by this bundle to a floating probe which is placed in the culture the turbidity of which one wishes to measure. The light passes through the culture and is reflected by a mirror, situated a short distance

*Vitatron Scientific BV, Kanaalweg 24, PO Box 76, NL 6210 Dieren, The Netherlands.

from the bottom of the bundle (but attached to it), and is then conducted by a second bundle of fibres to the photomultiplier of the spectrophotometer. The probe is shown in Fig. 22.

The particular device Robrish and his colleagues used was designed to fit the Gilford* 300N spectrophotometer, and Fig. 23 shows a growth curve obtained with the probe combined with that spectrophotometer. However, with minor modifications to that part of the unit which replaces the normal cuvette system, there seems to be no reason why this device should not be modified to suit a variety of spectrophotometers.

The principle of this technique is very different from the majority of others, in which either the culture vessel is placed in the instrument or the culture itself is drawn into the cell and ejected after the measurement. In this technique one is effectively taking the optics to the culture. Extending the optical path length in this way could present problems; it could, for example, alter the spectral response of the machine. In addition, of course, the probe has to be sterilized; the particular probe used by Robrish had to be sterilized by ethylene oxide. Nevertheless, the authors felt that a spectrophotometer equipped with such a fibre optic probe could be useful for conventional measurements, as well as for continuous culture work.

In its present form it may well not be convenient for use in processes

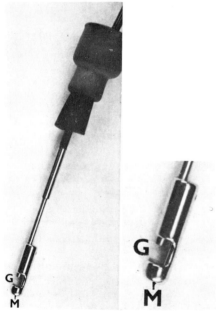

Fig. 22. The fibre optic turbidimetric probe devised by Robrish and his colleagues. Light is transmitted by the fibre optic to the top of the gap, G, across the gap, in which the culture is situated, and is reflected by a mirror in that part of the housing marked M. (From Robrish, S. A. et al. (1971). Use of a fibre optic probe for spectral measurements and the continuous recording of the turbidity of growing microbial cultures. *Appl. Microbiol.* **21**, 278–287.) (Photograph courtesy of Dr S. A. Robrish and the American Society for Microbiology.)

*Gilford Instruments Laboratories, 132 Artine Street, Oberlin, Ohio 44074, USA.

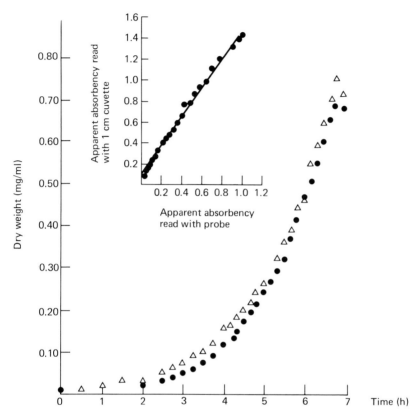

Fig. 23. Growth curves of *Streptococcus mutans* obtained with the fibre optic probe devised by Robrish and his colleagues, used in conjunction with a Gilford 300N spectrophotometer, and with a Gilford 2400 spectrophotometer used in the conventional way. (From Robrish, S. A. et al. (1971). Use of a fibre optic probe for spectral measurements and the continuous recording of the turbidity of growing microbial cultures. *Appl. Microbiol.* **21**, 278–287.) (Graphs courtesy of Dr S. A. Robrish and the American Society for Microbiology.)

frequently used in many routine laboratories; for example, in the measurement of the susceptibility of bacteria to antibiotics. Nevertheless, this technique could well form the basis of an extremely useful device for that purpose. It particularly lends itself to multichannel applications.

More sophisticated photometric devices

A sensitive, but relatively complicated, differential photometric turbidimeter was described by Piccialli and Piscitelli (1973). This is said to be particularly suitable for concentrations of organisms between 10^4 and 10^6 cells/ml.

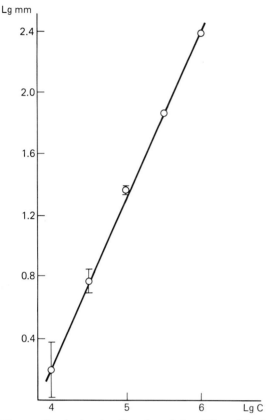

Fig. 24. Graph showing linearity of the differential photometric turbidimeter designed by Piccialli and Piscitelli (in concentration (C) range 10^4 to 10^6 *Staph. aureus*/ml). (From Piccialli,A. and Piscitelli, S. (1973). A simple turbidimeter for rapid determination of low bacteria concentrations. *Rev. Sci. Instrum.* **44,** 1717.) (Graph courtesy of Dr A. Piccialli and the American Institute of Physics.)

Fig. 25. The differential photometric turbidimeter designed by Piccialli and Piscitelli. (Photograph courtesy of Dr A. Piccialli.)

The light source is collimated and passed to two beam splitters, which divide the beam into three parts. One part, constituted of 6% of the total illumination, is used for the automatic control of the lamp. The other two parts, of equal intensities, are passed through reference and measuring cuvettes respectively. The result is automatically calculated and printed out on a recorder. Further details were not given.

Staphylococcus aureus suspended in a Merthiolate solution was used to measure the sensitivity of the instrument. Decreasing bacterial concentrations (which were measured by counting under a microscope) were prepared from the suspension, and three readings were made on each of 6 concentrations in the range from 3×10^4 to 18×10^4 cells/ml. The instrument could easily detect and resolve readings at 3×10^4 cells/ml, 6×10^4 cells/ml and so on. The linearity of the instrument in the range from 10^4 to 10^6 cells/ml was also tested, and is as seen in Fig. 24. The apparatus is shown in Fig. 25.

The authors indicate that this is a practical method of measuring bacterial concentrations of the order of 10^4 to 10^6 cells/ml. There are, however, certain details that one has to attend to because of the relatively high sensitivity of the instrument compared with the sensitivity of many other similar instruments. The liquid medium in which the organisms are counted has to be filtered and, to quote the authors, protected from contamination. Furthermore, between readings the cuvette has to be washed with filtered water and the recorder zero checked. It would appear that if one takes care with these details, the apparatus can be made to work, although there is some variation in the readings at the lowest concentration of the range quoted. The question is just how much trouble does one have to go to to obtain consistent results?

Another turbidimetric device, which utilizes an autoclavable, plastic 5 ml syringe as a cuvette, was devised by Fujita and Nunomura (1968). In this device (Fig. 26) an optically flat glass plate (G) is attached to the end of both the barrel (B) and plunger of the syringe. An optical rod (O), 6 mm diameter × 65 mm length, is placed in the plunger. A light beam (L) is passed through the glass plate in the barrel, through the culture which has been drawn into the syringe and along the length of the optical rod in the plunger to a photocell (P) which is attached to the outer end of the optical rod. The path length in the cuvette may be continuously varied in the range from 0.01 to 20.00 mm, by means of a micrometer which is attached to the plunger for this purpose, and the syringe is filled in the conventional way, through a 1

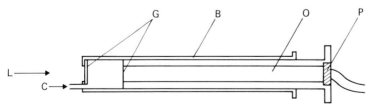

Fig. 26. The principle of the turbidimetric device devised by Fujita and Nunomura. L: light from source containing reference cell. C: culture inlet tube. G: optically flat glass plates. B: syringe barrel. O: optical rod in centre of syringe plunger. P: photocell.

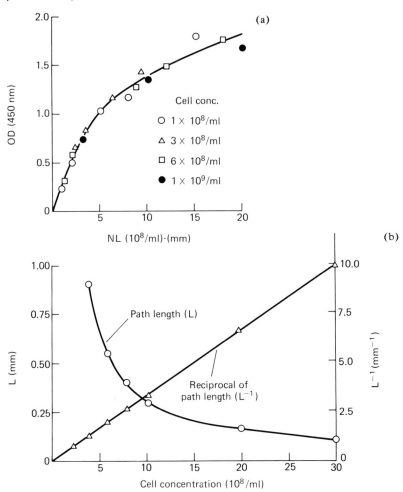

Fig. 27. Calibration curves of turbidimetric device devised by Fujita and Nunomura. (a) Curves obtained by plotting optical density (OD) vs product of cell concentration, N, and path length, L. (b) Curves obtained by plotting cell concentration vs path length and vs the reciprocal of path length. (From Fujita, T. and Nunomura, K. (1968). New turbidimetric device for measuring cell concentrations in thick microbial suspensions. *Appl. Microbiol.* **16**, 212–215.) (Graphs courtesy of Dr T. Fujita and the American Society for Microbiology.)

mm internal diameter needle mounted in the highest point of the glass plate attached to the end of the barrel. A microscope lamp is used as the light source, and part of the incident light is reflected from a half mirror to a reference photocell.

The calibration procedure is as follows. Suspensions of the organism to be tested in a variety of concentrations are made and the optical density for each concentration plotted against path length. These data enable one to plot a curve of optical density vs the product of cell concentration and path length (Fig. 27a). It was found with yeast suspension that with an optical density $\leqslant$

1, all the points fell on the same curve. For an optical density > 1, this was not the case. An alternative and more convenient method is to plot that path length (or the reciprocal of that path length) that gives the same optical density for each concentration vs cell concentration (Fig. 27b). With the aid of these curves one can obtain the concentration of an unknown specimen, simply by adjusting the path length to give a constant optical density and then reading off the concentration from the calibration curve. This procedure has to be conducted for every type of organism to be used in the machine.

This apparatus was devised specially for measuring the concentration of organisms in what are described as thick microbial suspensions. The authors used bakers yeast as a test microbe together with *Escherichia coli, Bacillus subtilis* and *Saccharomyces cerevisiae*. They found that the instrument was able to measure cell concentrations in the range from 10^6 to 10^9/ml for yeast and from 10^8 to 10^{11}/ml for bacteria.

This would appear to be a valuable method for measuring concentrations of yeast and organisms in the ranges quoted above. Whilst the process of calibrating the equipment is rather tedious, the operating procedure is very simple; one simply turns a micrometer until the optical density reaches a specific predetermined value. It is basically a very simple device to operate.

A device which has become available recently was described by Praglin et al. (1975). Devised specifically for measuring antibiotic sensitivities, this system is known as the Autobac 1 (Pfizer Diagnostics*).

The essence of this equipment is a disposable 13-chamber cuvette (Fig. 28) into 12 chambers of which antibiotic 'elution discs' are dispensed. The culture is dispensed into all 13 chambers by screwing the special bottle containing the culture into a holder on the cuvette and inverting it. After incubation for approximately 3 h at 35 deg C, during which the cuvette is shaken, it is removed and placed in the photometer unit which measures, in turn, light scattered at an angle of 35° from each of the chambers. The result, given simply as either sensitive, resistant or intermediate, is calculated automatically and printed out on a preinserted form.

McKie et al. (1975) carried out feasibility studies on this equipment. In this paper, they sought to justify the use of a fixed angle of 35° scatter and, also, the use of unpolarized light from a tungsten halide lamp. A variety of studies were reported. The growths of *Escherichia coli* and *Streptococcus pyogenes*, in different media, were measured and it was concluded that Eugonic broth was the most suitable. They reported experiments to determine the ideal initial inoculum concentration and incubation time; these were found to be 10^6 organisms/ml and 3 h respectively. We are here, of course, thinking solely in terms of measuring antibiotic sensitivities using the elution disc technique as described.

Although this machine was devised for measuring antibiotic sensitivities, clearly it could be used for other purposes since it is, in effect, a multichannel photometer. McKie and his colleagues are clearly satisfied that the Autobac 1 produces results for antibiotic sensitivities which compare well with the disc diffusion and MIC methods. No doubt, independent evaluation will be published in due time (see Johnston and Newsom, 1976). Similarly, the use

* Pfizer Diagnostics Division, 235 East 42nd Street, New York, NY 10017, USA.

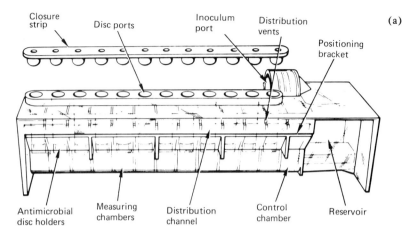

(a)

Closure strip
Disc ports
Inoculum port
Distribution vents
Positioning bracket

Antimicrobial disc holders
Measuring chambers
Distribution channel
Control chamber
Reservoir

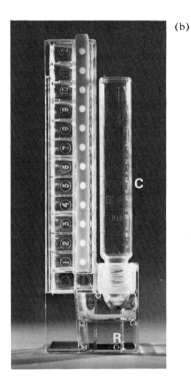

(b)

Fig. 28. The 13-chamber cuvette of the Autobac for measuring antibiotic sensitivities. (a) Drawing of the cuvette. (b) Photograph of cuvette with culture bottle attached. In the position shown, the reservoir, R, in the cuvette is filled with culture from bottle, C. The 13-cuvette chambers are filled by rotating the cuvette 90 degrees in an anti-clockwise direction. (Photographs courtesy of Dr J. Praglin and Pfizer Inc.)

of the Autobac 1 basic system for other microbiological measurements will doubtless be reported since it is, after all, very similar in principle to the devices produced by Cobb et al. (1970), Mackintosh et al. (1973) and Bergan (1975).

Another system using a disposable plastic cuvette has just become available. This is the Auto Microbic System (AMS)™ (McDonnell-Douglas*). Freeze-dried selective media are placed in the 20-well cuvette, and after inoculation, by means of a pneumatically controlled filling module, the cuvette is loaded into an incubator. The optical absorption is read, automatically, hourly for up to 12 hours, and the results are printed out.

The system may be used for microbial detection and identification, and the early work has concentrated on analysing urine specimens (Sonnenwirth, 1976). Further evaluations are in progress.

A series of papers by R. M. Berkman and P. J. Wyatt describe the use of what is effectively a nephelometer in which the detector measuring the light scattered from the culture in the cuvette can be moved through all angles between 5 and 175° with respect to the incident (laser) beam, so that a scattering characteristic for the culture can be obtained.

In 1968 Wyatt suggested that, because of the differences of structure in different types of organisms, the scattering characteristics would differ, and

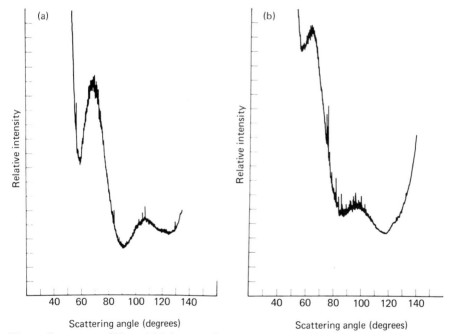

Fig. 29. Some recent differential light-scattering patterns obtained by the method of Wyatt and his colleagues. (a) A *Klebsiella* suspension. (b) *Proteus mirabilis*. (Photographic slides courtesy of Dr P. J. Wyatt and Science Spectrum Inc.)

* McDonnell-Douglas Astronautics Co., East St, St Louis, Missouri, USA.

thus provide a means for identifying organisms, and in 1969 he published some photographs of patterns obtained from *Streptococcus lactis* and *Serratia marcescens* in distilled water suspensions. Some more recent scattering patterns are shown in Fig. 29.

In 1970, Berkman et al. discussed the possibility of using their equipment, known as the Differential I (Science Spectrum Inc.*), for the rapid detection of penicillin sensitivity of *Staphylococcus aureus*, the thesis being that the effect of penicillin on the cell wall of a strain such as *Staph. aureus* would drastically alter the scattering pattern; they published curves (Fig. 30) showing the effects. Clearly, there are obvious differences in the scattering patterns. In this case the incident light was produced by a vertically polarized helium-neon laser.

This work led to the development of a machine specifically for measuring antibiotic sensitivities, known as the Differential III. This is based on exactly the same principle as the Differential I, but there are a variety of differences.

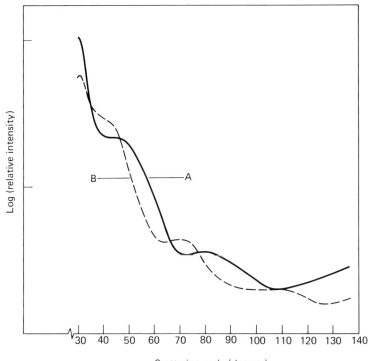

Fig. 30. Some recent curves showing the effects of penicillin on the differential light-scattering patterns of *Staph. aureus*. A: control culture. B: culture +0.14 units/ml penicillin. (From Wyatt, P. J. (1975). Automation of differential light scattering for antibiotic susceptibility testing. In: *Automation in Microbiology and Immunology*. Eds. C-G Hedén and T. Illeni. John Wiley.) (Courtesy Dr P. J. Wyatt, Science Spectrum Inc. and the Publisher.)

* Science Spectrum Inc., 1216 State Street, PO Box 3003, Santa Barbara, California 93105, USA.

A circular tray suitable for holding up to 10 cuvettes replaces the single cuvette of the earlier machine. The direct writing recorder on which the scattering characteristics are recorded has been replaced, because in the Differential III light-scattering patterns obtained with and without the antibiotic are compared automatically and the result is printed out on a scale calibrated between 0 and 99, 0 corresponding to resistance and 99 to sensitivity.

Berkman and his colleagues suggested that the differential light-scattering technique might well be useful for other drug studies. However, Thornsberry and Balows (1974) reported a thorough evaluation of the Differential I. They tested it with a variety of species of bacteria, *Escherichia coli, Proteus* sp., *Klebsiella pneumoniae, Enterobacter* sp. and alpha-haemolytic streptococci and a variety of antimicrobial agents, streptomycin, neomycin, gentamicin, kanamycin, lincomycin, carbenicillin and colistin. In summary, they found they were quite unable to reproduce the results obtained by Berkman and his colleagues. They found that the Differential I light-scattering machine could not be used for clinically useful susceptibility studies. They did not test the Diffential III, but since the modifications of the Differential I that have been built into the Differential III are in no way connected with the basic principle of measuring and of interpreting the differential light-scattering characteristics, but are merely matters of convenience of handling the number of cultures required and of presenting the results in the way required for antibiotic sensitivity testing, one cannot help but have doubts about the suitability for that purpose. However, Wyatt maintained his confidence in this technique, as shown in a recent paper (Wyatt, 1975).

It appears that photometry is much more likely to become a valuable diagnostic technique for measuring concentrations of bacteria in suspension than other techniques, particularly electronic cell-counting techniques (see p. 52). Many of the technical problems have been solved and one does not necessarily have to handle infected material directly.

A wide range of devices have been produced and a few designs specifically for bacteriological applications have become commercially available. Unfortunately, these are generally rather complex and, as we have seen, yet to be proven. I would like to see more effort devoted to devising simpler devices specifically for bacteriological procedures because there is clearly a pressing need for such effort. A word of caution, however, because such devices have their limitations — one of the biggest of which is that generally their sensitivity is relatively poor; the minimum concentration of organisms in suspension that can be detected is usually of the order of 10^6 organisms/ml. The electronic capillary tube devices discussed earlier are more sensitive.

Cell counting

I do not intend to dwell long on this topic since, as already suggested, photometric techniques are more likely to become a valuable diagnostic technique for measuring concentrations of bacteria in suspension than cell counting techniques. Nevertheless, cell counters, particularly the Coulter counter, have been used in bacteriology, so brief mention is made merely to clarify and complete the discussion on the enumeration of bacteria in suspension.

The Coulter* cell counter (Coulter, 1953) is widely used in haematology for counting red blood cells (Brecher et al., 1956) and white blood cells (Richar and Breakell, 1959). In principle, a sample is diluted in an electrically conducting fluid (usually saline) and the beaker containing the mixture is placed on a platform so that a probe is immersed in the mixture; the probe consists of a flat plate electrode and a glass tube enclosing a second flat plate electrode. There is a small orifice in the wall of the glass tube and an electric potential is applied across the electrodes, thus producing an electric field across the orifice. The diluted sample is drawn through the orifice and, as each blood cell passes through, it displaces some of the conductive fluid thus raising the impedance of the orifice contents and producing a voltage pulse. The amplitude of the pulse is proportional to cell size. The output signal may be displayed on an oscilloscope or it may be processed to give the total count in digital form. Pulse amplitude discrimination techniques may be used to count cells within a predetermined size range, and size distribution curves can be obtained in a few minutes.

The Coulter counter was first used for counting bacteria in suspension by Kubitschek (1958). He modified the standard cell counter, by replacing the 100 μm orifice with a 10 μm orifice, and counted *Escherichia coli* (strain B) and *Bacillus megaterium* spores; 4-hour cultures of each were diluted in 0.1 mol/l hydrochloric acid. To calibrate the apparatus, he suspended 1.13 μm and 3.2 μm polystyrene latex spheres in saline and counted them on a haemocytometer to obtain an absolute count and compared the count with that obtained on the electronic counter. Agreement to within 2% was obtained. Kubitschek also used the electronic counter for sizing the same organisms, but he gave very little information about the technique; he simply stated that he used a pulse height analyser to give the cell volumes.

A more detailed investigation was carried out by Curby et al. (1963). They grew *Staphylococcus aureus* (SM), *Escherichia coli* (SIAS) and an *E. coli* variant and *E. freundii* (8454) and an *E. freundii* variant in brain-heart infusion broth. Cultures inoculated from an agar slant and from a broth culture were diluted 0.9% in sodium chloride solution and counted on a model A Coulter counter, with a 30 μm orifice, at 1 h intervals after inoculation; the inoculation techniques were not discussed. Similarly, pour plates were made of each culture. Curby and his co-workers found that the electronic count depended on the magnitude of the electric field applied across the orifice of the counter and that the relationship between counts and electric field depended on the organism. The phenomenon was not investigated in detail but the authors concluded that 'it was not related to a charge on the organism but was determined by an active process within the organism just before cell division.' In general, the electronic count obtained for the *E. coli* and its variant agreed with the pour plate method for the culture inoculated from an agar slant but poor agreement was obtained for the culture inoculated from a broth culture. With the *E. freundii* and its variant poor agreement was obtained for both cultures and with the *Staph. aureus* the electronic count was always higher than the pour plate count.

---

* Coulter Electronics Inc., 590 West 20th Street, Hialeah, Florida 33010, USA.

Since that time, several workers have used the machine for counting and for sizing bacteria. For example, Manor and Haselkorn (1967) used the cell counter for the size fractionization of exponentially growing *E. coli*; in principle, different size fractions of bacteria in a culture were prepared in a zone centrifuge and a sample of each fraction was counted to give information about the age of the organisms in the culture. Andrew and Westwood (1971) used the counter to count cells of *Streptococcus faecalis*; they found that growth curves produced from the counter were reproducible and similar to the shape of curves produced by other methods, such as viable pour plate counts, total counts and nephelometry. McCarthy (1971) used the counter for growth studies of *Mycobacterium avium*, and Anderson and Whitehead (1973) counted *Escherichia coli*, *Salmonella typhimurium* and an *Enterobacter* species from commercially manufactured ice-cream.

Whilst, as we have seen from the brief discussion above, the Coulter counter has been used for counting and sizing bacteria, the success has been very limited; there are doubtless a variety of reasons for this. The counter cannot distinguish live bacteria from dead bacteria, or bacteria from other particles which may be in the suspension, and consequently the total viable count obtained by the pour plate method, which is taken to be the number of viable colony-forming units, could well be lower than the electronic count. Also the electronic counter cannot distinguish single bacteria from aggregates and this is likely to cause errors that offset errors due to the aforementioned factors. (When counting colonies on pour plates one has to be cautious of aggregates, but they are usually clearly recognizable.) The effects due to all these factors must vary considerably from specimen to specimen and it is likely to prove difficult to formulate general criteria for relating the electronic count to the total viable count.

The evidence indicates that much more work needs to be carried out before the Coulter counter or any similar apparatus, such as the Celloscope,* can be used as a diagnostic aid, although it appears it is possible to count bacteria under ideal conditions, using an electrolyte with few particles in it and growing the organism in a specially prepared particle-free medium. One might, therefore, be able to use cell counting techniques for the measurement of antibiotic sensitivities although one can visualize many problems. The difficulties of using the machine to count organisms in a clinical specimen such as urine are considerable and have yet to be overcome.

References

Colony counters
For counting colonies in capillary tubes

Blume, P., Johnson, J. W. and Matsen, J. M. (1975). Automated antibiotic susceptibility testing. In: *Automation in Microbiology and Immunology.* Eds. C-G. Hedén and T. Illeni. John Wiley, New York and Chichester.

Bowman, R. L., Blume, P. and Vurek, G. G. (1967). Capillary-tube scanner for mechanized microbiology. *Science* **158**, 78.

* Linsom Instruments, Osthammarsgatan 80, S-11528 Stockholm, Sweden.

Hartman, P. A. (1968). *Miniaturized Microbiological Methods.* Academic Press, New York and London.

Nakamura, H., Samejima, K. and Tamura, Z. (1974). A capillary tube method for counting viable cells of *Bifidobacterium bifidum* grown in a solid medium. *Jap. J. Microbiol.* **18**, 135.

Schoon, D. J., Drake, J. F., Fredrickson, A. G. and Tsuchiya, H. M. (1970). Automated counting of microbial colonies. *Appl. Microbiol.* **20**, 815.

Yanagita, T. (1956). Capillary tube method for counting viable bacteria. *J. Bact.* **71**, 381.

For counting surface colonies

Alexander, N. E. and Glick, D. P. (1958). Automatic counting of bacterial cultures — a new machine. *I.R.E. Trans. Med. Electron. PGME* **12**, 89.

le Bouffant, L. and Soule, J. L. (1954). The automatic size analysis of dust deposits by means of an illuminated slit. *Brit. J. appl. Phys.* Suppl. 3, S143.

Goebel, W. F. and Blum, J. (1950). A bacteriophage plaque and colony counter. *J. exp. Med.* **92**, 541.

Goss, W. A., Michaud, R. N. and McGrath, M. B. (1974). Evaluation of an automated colony counter. *Appl. Microbiol.* **27**, 264.

Malligo, J. E. (1965). Evaluation of an automatic electronic device for counting bacterial colonies. *Appl. Microbiol.* **13**, 931.

Miles, A. A. and Misra, S. S. (1938). The estimation of the bactericidal power of the blood. *J. Hyg. (Lond.)* **38**, 733.

Moore, W. T. and Taylor, C. B. (1950). An improved colony illuminator. *J. gen. Microbiol.* **4**, 448.

Murphy, R. P. and Tucker, C. G. (1970). A comparator method of assessing bacterial plate counts. *J. appl. Bact.* **33**, 641.

Yourassowsky, E. and Schoutens, E. (1974). Automated count and size evaluation of colonies of bacteria grown in a zonal concentration gradient of antimicrobial agent. *Appl. Microbiol.* **28**, 525.

Yourassowsky, E. and Schoutens, E. (1975). Automated counting of colonies by Micro-Videomat. In: *Automation in Microbiology and Immunology.* Eds. C-G. Hedén and T. Illeni. John Wiley, New York and Chichester.

Other colony counting techniques

Briner, W. W., Wunder, J. A. and Blair, D. Q. (1975). Use of rapid bacterial plating and counting techniques in a study of the microbial ecology of skin. Paper read at Int. Conf. on Mechanized Microbiology, Ottawa, Canada, Sept. 1975.

Campbell, J. E. (1976). Estimation of microbial density through pattern recognition. In: *Proceedings 2nd International Symposium on Rapid Methods and Automation in Microbiology.* Eds. H. H. Johnston and S. W. B. Newsom. Learned Information (Europe) Ltd., Oxford.

Campbell, J. E. and Gilchrist, J. E. (1973). Spiral plating technique for counting bacteria in milk and other foods. *Dev. Industr. Microbiol.* **14**, 95.

Gilchrist, J. E., Campbell, J. E., Donnelly, C. B., Peeler, J. T. and Delaney, J. M. (1973). Spiral plate method for bacterial determination. *Appl. Microbiol.* **25**, 244.

Read, Jr., R. B., Gilchrist, J. E., Donnelly, C. B. and Campbell, J. E. (1974). The spiral system for plating and counting bacteria. Abstract of the 9th Symposium of International Assn. Microbiology Societies, Kiel, Germany, Sept. 1974.

Sharpe, A. N. (1973). Automation and instrumentation developments for the bacteriology laboratory. In: *Sampling — Microbiological Monitoring of Environments.* Eds. R. G. Board and D. W. Lovelock. Academic Press, New York and London.

Sharpe, A. N. and Kilsby, D. C. (1971). A rapid, inexpensive bacterial count technique using agar droplets. *J. appl. Bact.* **34**, 435.

Measuring concentrations of bacteria in suspension

Simple photometric devices

Alper, T. and Sterne, M. (1933). The measurement of the opacity of bacterial cultures with a photo-electric cell. *J. Hyg. (Lond.)* **33**, 497.

Bergan, T. (1975). Automatic turbidimetric recorder for microbial cultures. In: *Automation in Microbiology and Immunology.* Eds. C-G. Hedén and T. Illeni. John Wiley, New York and Chichester.

Cobb, R., Crawley, D. F. C., Croshaw, B., Hale, L .J., Healey, D. R., Pay, F. J., Spicer, A. B. and Spooner, D. F. (1970). The application of some automation and data handling techniques to the evaluation of antimicrobial agents. In: *Automation, Mechanization and Data Handling in Microbiology.* Eds. A. Baillie and R. J. Gilbert. Academic Press, New York and London.

Kavanagh, F. (1963). *Analytical Microbiology.* Academic Press New York and London.

Mackintosh, I. P., Watson, B. W. and O'Grady, F. W. (1973). Development and further applications of a simple turbidity cell for continuously monitoring bacterial growth. *Phys. in Med. Biol.* **18**, 265.

Meynell, G. C. and Meynell, E. (1965). *Theory and Practice in Experimental Bacteriology.* Cambridge University Press, Cambridge.

Norris, K. P. (1959). Infra-red spectroscopy and its application to microbiology. *J. Hyg. (Lond.)* **57**, 326.

Norris, J. R., Hewett, A. J. W., Kingham, W. H. and Perry, P. C. B. (1970). An automatic growth recorder for microbial cultures. In: *Automation, Mechanization and Data Handling in Microbiology.* Eds. A. Baillie and R. J. Gilbert. Academic Press, New York and London.

Pulvertaft, R. J. V. and Lemon, C. G. (1933). Application of photo-electricity to the determination of bacterial growth rate. *J. Hyg. (Lond.)* **33**, 245.

Robrish, S. A., LeRoy, A. F., Chassy, B. M., Wilson, J. J. and Krichevsky, M. I. (1971). Use of a fiber optic probe for spectral measurements and the continuous recording of the turbidity of growing microbial cultures. *Appl. Microbiol.* **21**, 278.

Snell, F. D. and Snell, C. T. (1948). *Colorimetric Methods of Analaysis.* D. Van Nostrand, New York and Wokingham, Berks.

Tennant, G. B. and Withey, J. L. (1970). A simple high-speed recording colorimeter system. *J. clin. Path.* **23**, 452.

Watson, B. W., Gauci, C. L., Blache, L. and O'Grady, F. W. (1969). A simple turbidity cell for continuously monitoring the growth of bacteria. *Phys. in Med. Biol.* **14**, 555.

More sophisticated photometric devices

Bergan, T. (1975). Automatic turbidimetric recorder for microbial cultures. In: *Automation in Microbiology and Immunology.* Eds. C-G. Hedén and T. Illeni. John Wiley, New York and Chichester.

Berkman, R. M., Wyatt, P. J. and Phillips, D. T. (1970). Rapid detection of penicillin sensitivity in *Staphylococcus aureus. Nature (Lond.)* **228**, 458.

Cobb, R., Crawley, D. F. C., Croshaw, B., Hale, L. J., Healey, D. R., Pay, F. J., Spicer, A. B. and Spooner, D. F. (1970). The application of some automation and data handling techniques to the evaluation of antimicrobial agents. In: *Automation, Mechanization and Data Handling in Microbiology.* Eds. A. Baillie and R. J. Gilbert. Academic Press, New York and London.

Fujita, T. and Nunomura, K. (1968). New turbidimetric device for measuring cell concentrations in thick microbial suspensions. *Appl. Microbiol.* **16**, 212.

Johnston, H. H. and Newsom, S. W. B. (Eds.) (1976). *Proceedings 2nd International Symposium on Rapid Methods and Automation in Microbiology.* Learned Information (Europe) Ltd., Oxford.

Mackintosh, I. P., Watson, B. W. and O'Grady, F. (1973). Development and further applications of a simple turbidity cell for continuously monitoring bacterial growth. *Phys. in Med. Biol.* **18**, 265.

McKie, J. E., Borovoy, R. J., Dooley, J. F., Evanega, G. R., Mendoza, G., Meyer, F., Moody, M., Packer, D. E., Praglin, J. and Smith, H. (1975). Autobac 1 — a 3-hour, automated antimicrobial susceptibility system; II. Microbiological Studies. In: *Automation in Microbiology and Immunology.* Eds. C-G. Hedén and T. Illeni. John Wiley, New York and Chichester.

Piccialli, A. and Piscitelli, S. (1973). A simple turbidimeter for rapid determination of low bacteria concentrations. *Rev. Sci. Instrum.* **44**, 1717.

Praglin, J., Curtiss, A. C., Longhenry, D. K. and McKie Jr., J. E. (1975). Autobac 1 — a 3-hour, automated antimicrobial susceptibility system; I. System description. In: *Automation in Microbiology and Immunology.* Eds. C-G. Hedén and T. Illeni. John Wiley, New York and Chichester.

Sonnenwirth, A. C. (1976). Developmental studies of an automated microbial detection and identification system. In: *Proceedings 2nd International Symposium on Rapid Methods and Automation in Microbiology.* Eds. H. H. Johnston and S. W. B. Newsom. Learned Information (Europe) Ltd., Oxford.

Thornsberry, C. and Balows, A. (1974). Automation of antimicrobial susceptibility testing. In: *Current Techniques for Antibiotic Susceptibility Testing.* Ed. A. Balows. Charles C. Thomas, Springfield, Illinois.

Wyatt, P. J. (1968). Differential light scattering: a physical method for identifying bacterial cells. *Appl. Optics* **7**, 1879.

Wyatt, P. J. (1969). Identification of bacteria by differential light scattering. *Nature (Lond.)* **221**, 1257.

Wyatt, P. J. (1975). Automation of differential light scattering for antibiotic susceptibility testing. In: *Automation in Microbiology and Immunology.* Eds. C-G. Hedén and T. Illeni. John Wiley, New York and Chichester.

Cell counting

Anderson, G. E. and Whitehead, J. A. (1973). Viable cell and electronic particle count. *J. appl. Bact.* **36**, 353.

Andrew, M. H. E. and Westwood, N. (1971). Use of the Coulter counter to count cells of *Streptococcus faecalis. J. appl. Bact.* **34**, 441.

Brecher, G., Schneiderman, M. and Williams, G. E. (1956). Evaluation of electronic red blood cell counter. *Amer. J. clin. Path.* **26**, 1439.

Coulter, W. H. (1953). US Patent No. 2, 656, 508.

Curby, W. A., Swanton, E. M. and Lind, H. E. (1963). Electrical counting characteristics of several equivolume micro-organisms. *J. gen. Microbiol.* **32**, 33.

Kubitschek, H. E. (1958). Electronic counting and sizing of bacteria. *Nature (Lond.)* **182**, 234.

McCarthy, C. (1971). Electronic counting in growth studies of *Mycobacterium avium. Appl. Microbiol.* **22**, 546.

Manor, H. and Haselkorn, R. (1967). Size fractionation of exponentially growing *Escherichia coli. Nature (Lond.)* **214**, 983.

Richar, W. J. and Breakell, E. S. (1959). Evaluation of an electronic particle counter for the counting of white blood cells. *Amer. J. clin. Path.* **31**, 384.

3
Devices for distributing chemically clean fluids

All the devices to which I referred in Chapter 1 — for distributing infected and/or sterile fluids — may also be used for distributing fluids that have to be clean but not sterile, as required, for example, in serological and similar procedures. In this chapter we consider the applications of some of those devices to such procedures.

In addition, we discuss some devices that have been devised primarily for handling clean fluids, but it does not follow that they invariably cannot be used for handling sterile fluids. It will become evident that whether a device can be used for handling sterile fluids or not partially depends on the suitability of chemical sterilization for the particular application and fluid one has in mind. It is not appropriate to pursue the efficacy of chemical or other sterilization procedures here.

Pipetting/transferring small volumes of fluid

Diluting loops

We have already considered the Takatsy loops and their use for, in particular, the measurement of antibiotic sensitivities. However, many of the early applications of these devices were for performing serological titrations, and there has been substantial growth in the development of microtitre equipment for this purpose. A few of the applications are given below.

Sever and his colleagues (Fuccillo et al., 1970) used the spiral loops for serially diluting serum in a specific microindirect haemagglutination test (HA) for *Herpesvirus hominis* Types I and II, which test was a development of the agglutination test described by Scott et al. (1957), and compared the results of their tests with those obtained by the microneutralization and standard complement fixation (CF) tests; the CF tests also utilized the microtitre spiral loops.

Paired human sera from 14 patients with rising titre of complement-fixing antibodies to herpesvirus were tested by all three methods. The type specificity of the H antibody response obtained by the HA test correlated well with the data obtained with the microneutralization test. The HA test was found to be rapid and simple for the detection of Type I and Type II herpesvirus infections.

Oberhofer and Hajkowski (1970) used microtitre loops for heterophilic antibody determinations. They tested 233 sera, sent to the laboratory for examination for infectious mononucleosis, by the microtitre technique and also by the standard tube dilution test.

Into each well of a microtitre tray was dispensed 25 µl of saline, and 25 µl of diluted inactivated serum was mixed with the saline in the first well of each row. Eleven twofold dilutions were made with the 25 µl loop, sheep cells were added and the plates incubated for approximately 2 h. The standard tube test was performed on 215 of the specimens. They found that in approximately 51 % of the tests both procedures gave identical results. In approximately 44 % of the cases there was a discrepancy of 1 dilution and in approximately 6 % of the cases there was a discrepancy of 2 dilutions.

All 233 microtitre tests were repeated. In 151 of the sera identical findings were obtained and another 78 sera gave differences that were no greater than 1 dilution, giving an overall reproducibility, if one assumes a 1 dilution difference is acceptable, of 98 %.

The authors indicated that there was a 'moderate' saving in the cost of reagents used in their tests, but they regarded the rapidity of performing the micro test as of greater significance; they found a technician could perform 5 or more complete micro tests in one-third of the time required for performing a similar number of macro tests.

On 28 specimens they also carried out studies comparing procedures using 25 µl volumes of reagents with procedures using 50 µl volumes. They found that using 50 µl volumes made the tests more difficult to read, because the larger volume of cells required a longer time for settling. They also found that the 25 µl test gave better correlation with the standard tube test than the 50 µl test gave.

O'Brien et al. (1971) investigated the reproducibility of the haemagglutination (HA) and haemagglutination-inhibition (HI) tests for influenza virus, by comparing the standard tube test with micro techniques in which the antigens and sera respectively were serially diluted by means of spiral loops which were operated manually and also semi-automatically, as in the equipment (see p. 9) described by MacLowry and Marsh (1968).

The loops, whether used manually or semi-automatically, were always presoaked in the same diluent as that used in the test. The dilutions were performed by rotating 8 loops simultaneously and all reagents were delivered by hand, although brief reference was made to an 'Autopipettor'. We shall return to that machine later (p. 63).

In the HA test it was found, by statistical analysis, that the semi-automatic microtitre procedure was more reproducible than the standard tube method by a factor of 1.84. The manual microtitre procedure was the least reproducible. In the HI test, the standard tube method was found to be the most reproducible, but only small differences between the manual and semi-automatic microtitre methods and the tube method were observed. The authors pointed out that although one saves a great deal of time and reagents by using the microtitre technique, this might not in itself justify the use of it in place of the standard tube test. They feel there must be adequate statistical evidence of improved reproducibility to justify introducing this test. The semi-automatic micro technique was found to be highly reproducible for the determination of both HA and HI titres.

The fundamental accuracy of the microtitre technique, particularly in the haemagglutination-inhibition tests, was studied in considerable detail by

Ashcroft et al. (1971). They evaluated the two types of dilutor (see p. 6): that made of a coil of wire, to which they referred as the coil loop; and that made out of solid metal, to which they referred as the lotus loop. (They also evaluated the 25 µl dropping pipette; their results are given in the next section.) The loops were those supplied by the Cooke Engineering Company* and they were tested in two ways, by diluting a potassium iodide solution labelled with ^{131}I in unlabelled potassium iodide, and by diluting normal human serum iodinated with ^{131}I in a Borate buffer solution of pH 9 containing 0.4% bovine albumen.

In the former method the authors found that there was no significant difference in the performances of the coil and the lotus loops, both of which transferred, on average, 90% of the expected amount. There was no consistent bias. Each dilution was repeated 4 times and all 4 specimens were counted simultaneously.

In the second method, an average of 24 and 28 titrations with the coil and the lotus loops respectively were made. The authors found that both were very accurate but that the coil loop gave a consistent negative bias. This was so small, however, that there was 'little chance' that this would introduce as much as 1 dilution difference from the true result when performing a series of up to 8 dilutions.

Ashcroft and his colleagues made the interesting observation that, as one would expect, the coil loops tend to lose their shape over a period of use, but that in a series of 6 experiments carried out with old coil loops which were badly mis-shapen, the results obtained were very similar to those obtained with a set of new coil loops, which one might not expect. Nevertheless, they concluded that if one wishes to perform no more than 8 serial dilutions, the choice of dilutor would depend on durability and that therefore the lotus loop would probably be the preferred type. Although they do not say so, it would appear from the information given that with the coil loop an error, namely 1 dilution difference, would appear after 10 serial dilutions due to the consistent and negative bias found with this type of loop.

Microtitre equipment for use with 10 and 20 µl quantities of fluid has become available, and Fitzgerald et al. (1974) carried out comparative microtitrations with the 10 µl and 25 µl systems; the 25 µl system is now widely used throughout the world, and Fitzgerald regards it as the 'standard' system. Parallel duplicate tests were performed using the indirect haemagglutination test for cytomegalovirus, herpes Type I and Type II, and for the complement fixation test for mumps, but, unfortunately, only 6 sera were tested in the latter case.

In the indirect haemagglutination test, 48 sera were tested for herpes Type I virus and 37 gave an identical titre by both the 10 µl and the 25µl systems. The remaining 11 sera produced a discrepancy of one twofold dilution. Similarly, 48 sera were tested for herpes Type II virus and 36 specimens gave identical titres; 10 produced a discrepancy of one twofold dilution. A total of 10 sera were tested for cytomegalovirus and an identical titre was produced in all cases.

*Cooke Engineering Company, now incorporated in Dynatech Laboratories Inc., 900 Slaters Lane, Alexandria, Virginia 22314, USA.

Fitzgerald and her co-workers pointed out that the large-scale use of the 10 µl system may be tedious, due to the very much smaller size of the wells compared to the wells used in the 25 µl system. Unfortunately, the authors do not say whether the loops were operated manually or semi-automatically, but presumably it was manually.

This brief discussion would indicate that the 10 µl system is likely to be a practical proposition. However, the errors inherent in a system in which one is dealing with such small volumes of fluids are bound to be high. Fitzgerald herself indicated that there are difficulties when using the 10 µl pipette droppers to distribute the diluent in her system. It is very probable that one would have to take much more care with the equipment and with the procedures when using 10 µl volumes than one would when using 25 µl volumes or above. One cannot help but wonder whether the disadvantages outweigh the advantages in using such small volumes.

Hitherto, I have not discussed the plates in which the serological procedures were carried out, but as early as 1964 Sever and his colleagues recognized that the shape of the bottom of the wells, whether it be round or conical, could be important particularly in looking for agglutination. They found that complement fixation tests were best performed in plates with round-bottomed wells, whereas haemagglutination and haemagglutination-inhibition tests were most satisfactory when conducted in conical-bottomed wells.

Nevertheless, it would appear that it is very much a matter of personal choice and the ease with which any particular individual can read the plates. However, this does not necessarily follow; the shape of the cup, as well as its material, may affect the test and this is something that should be considered in devising techniques.

Pipettes

The microtitre diluting loops are not the only devices used for distributing infected and/or sterile fluids that may be used for distributing fluids for serological procedures, or for any other purpose, and as one would expect, the dropping pipette (and also the semi-automatic high precision pipette) has been used in a variety of other applications. For example, 50 µl and 25µl droppers were used by Reinisch and Hall (1966) to deliver sample, diluent and antigen in a technique to detect infectious mononucleosis. They appear to have assumed that the accuracy of these droppers quoted by the manufacturers is attained. However, they do indicate that the tip of the dropper must be carefully dried with absorbent material after aspiration, and that the dropper must be held vertically whilst drops are being delivered because otherwise larger drops than the nominal size drop may form.

The droppers manufactured by Cooke Engineering Company* were evaluated by Ashcroft *et al.* (1971), by weighing 10 drops of diluent containing 0.4% bovine albumen (the diluent consisted of a Borate buffer solution of pH9 containing 0.7% sodium chloride, 0.3% boric acid and 0.08% sodium hydroxide). The mean of the 10 weighings gave a value for the volume dispensed of 0.02569 ml/drop, with a coefficient of variation < 2%.

*Cooke Engineering Company, now incorporated in Dynatech Laboratories Inc., 900 Slaters Lane, Alexandria, Virginia 22314, USA.

Semi-automatic pipettes and similar devices

An example of the application of a semi-automatic pipette (see p. 3) is that of Williams and Whittemore (1971), who published details of a microagglutination technique for diagnosing pullorum disease (*Salmonella pullorum*): they used a 10µl Oxford* sampler to deliver serum into each well used in that technique. The tip was rapidly rinsed in fresh saline, by depressing the plunger and slowly releasing it to the starting position several times between each sample, to remove any trace of serum; the tip was then touched on to absorbent material, with the plunger at the lowest position, in order to remove all traces of saline before dispensing the next sample. They replaced the tip every 100 sera, but suggested that an unlimited number of deliveries can be made provided the tip is carefully rinsed between samples. They noted that hand fatigue can occur and suggested that semi-automatic pipettes are suitable only for delivering a limited number of sera; whether this is because of the hand fatigue or for other reasons is not clear. Unfortunately, they did not evaluate the performance of the pipetting devices separately, but evaluated only the performance of the complete microagglutination procedure.

Many devices for producing small drops of fluid have been devised, and a very common one is that which is in essence a 'Hamilton' precision syringe, the plunger of which is moved by a micrometer; a micrometer is used in order that the plunger of the syringe may be moved a precise and reproducible distance, thus producing a constant drop volume. An example of this was described by Wederkinch (1964). In this device, the syringe is filled, in the normal way, and then placed into the micrometer; the screw is turned so that a drop of fluid forms on the tip of the syringe needle, the drop is removed and the device is then ready for use. Wederkinch found, for example, that the machine would repeatedly dispense 400 µl and 200 µl volumes with coefficients of variation of 0.00020 and 0.00034 respectively.

Kropp (1963) devised a similar device which would deliver quantities of the order of 25 µl using a 1 ml syringe; this volume could be delivered to an accuracy of better than 2%.

A repeating dispenser, a device which will eject 1/50th of the capacity of a syringe at each press of a button, is marketed by Hamilton.† A number of syringes, with capacities in the range of 25 µl to 2.5 ml, will fit into the same holder so that volumes between 0.5 and 50 µl may be repeatedly dispensed. A larger device that will hold 5 ml and 10 ml syringes, thus enabling one to repeatedly dispense volumes of 0.1 ml and 0.2 ml respectively, is also available.

I know of no papers in which details of an evaluation of this device have been published, although such information may well have appeared incidentally in a report of work, for example, in dispensing antigens. Unfortunately, the manufacturers do not quote the accuracy with which it will dispense such small volumes as 0.5 µl. Clearly, one would have to find

---

*Oxford Laboratories (International), 1149 Chess Drive, Foster City, California, USA.
†Hamilton Company, Reno, Nevada, USA.

that out for oneself. Experience with a similar device, however, suggests that problems could arise through wear of the mechanism and of the syringe plunger, and that the performance generally deteriorates rapidly with use. One would need to be on guard for such an eventuality. One should note that high precision syringes have to be kept very clean (regardless of the device in which they are used) or the plunger will easily stick.

There must be hundreds of spring-loaded syringes, a syringe the plunger of which is attached to a framework which in turn is spring-loaded so that it automatically moves as far as is permitted by an adjustable screw or stop, in use throughout the world. In these devices, the volume that is sucked up and discharged depends on the length of plunger movement, and this volume is, therefore, adjustable simply by adjusting the position of a stop. The framework is so designed that it can easily be held in the hand and operated by a simple movement of the thumb (Fig. 31).

The design of these devices varies, and in some cases it is possible to use them with a detachable Pasteur or similar pipette—in which case, they can be used for handling infected fluids, provided the fluid concerned does not pass beyond the end of the sterilizable pipette. The accuracy of these devices varies; they are commonly used with volumes of the order of 1 ml or above and one would expect them to be accurate to $\pm 1\%$ when handling such volumes. Care has to be taken because the fluid is moved by a force which is transmitted through air, a highly compressible substance, and this reduces the accuracy considerably.

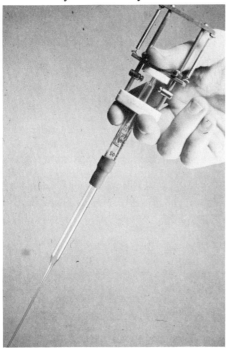

Fig. 31. A spring-loaded syringe.

Multiple pipettors

One type of multiple pipettor, which is part of the Autotitre apparatus (Ames*), has already been mentioned on p. 14. Unfortunately, detailed information was not given in the paper. (But the devices, used for dispensing diluent and inoculum in a method for measuring minimum inhibitory concentrations of antibiotics, can only be sterilized chemically and one wonders about the advisability of using pathogens in such apparatus.)

A device consisting of 12 vertically mounted syringes, the plungers of which are operated by a simple hydraulic system, was devised by Sanderson (1968, 1970): his apparatus was produced specifically for use with microtitre equipment for serological procedures. The 12 syringes—the dispensing syringes— are mounted in a straight line and the barrels (D) are attached to a fixed frame (F) under which the microtitre plates (M) are moved (Fig. 32). The plungers of the syringes (P) are attached to a bar (B), which slides on the frame. This bar is attached to the plunger (C) of another (larger) syringe—the control syringe—the barrel of which (S) is also fixed to the frame. Fluid is forced into this syringe, by means of a very simple hydraulic pump (an automatic pipette similar to the Cornwall pipette discussed on p. 65), so that the control syringe plunger (C) moves, thus moving, in turn, the bar (B) and the dispensing syringe plungers (P) to discharge a volume of fluid from each syringe: the volume of fluid forced into the control syringe is so arranged that 25 µl is discharged from each dispensing syringe.

As Sanderson indicated, there is a multitude of applications for this equipment. He used it for virus haemagglutination-inhibition, complement fixation, haemagglutination and antiglobulin tests. Unfortunately, he did not indicate the accuracy of his equipment. He did say, however, that with it one person could screen 100 sera against up to 11 viruses in 2 h working time.

An instrument, called the Autopipettor, was mentioned by O'Brien et al. (1971). This was designed for adding 25 µl of diluent or 50 µl of red-blood-cell suspension in their method for performing the haemagglutination (HA) and haemagglutination-inhibition (HI) tests using the microtitre system. Unfortunately, they do not describe the equipment or even indicate the number of channels. One assumes, however, that it is 8 × 12, and that the apparatus is the forerunner of similar apparatus manufactured by the American Instrument Company†. They simply say that the reproducibility of the HA and HI procedures can be improved by using the Autopipettor instead of distributing diluents and red blood cells by hand.

Williams and Whittemore (1971) very briefly referred to two devices for adding quantities of saline, antigen and other reagents to microwells in any microtitre procedure. These were a 96- and a 12-channel hand-held dispenser. They did not discuss how they functioned, except to say that in order to vary the volume delivered from each needle, one varies the total volume of fluid introduced into the device. Presumably, the total volume is divided into 96 (or 12) equal portions, but how or with what accuracy one is not told. They indicated that a Brewer automatic pipetting machine (an

*Miles Laboratories Inc., Ames Division, 1127 Myrtle Street, Elekhart, Indiana 46514, USA.
†American Instrument Company, 8030 Gerogia Avenue, Silver Spring, Maryland 20910, USA.

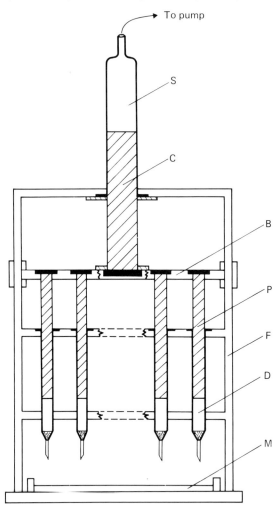

To pump

S

C

B

P

F

D

M

Fig. 32. The principle of the multiple dropper devised by Sanderson. S: control syringe barrel. C: control syringe plunger. B: bar, to which dispensing syringe plungers, P, are attached, which slides on fixed supporting frame, F. D: dispensing syringe barrels. M: microtitre plate.

automatically operated fluid-dispensing machine) and a Cornwall syringe (as discussed in the next section) have been used to supply fluids to these particular multiple dispensers.

The authors said that these devices were being examined with a view to their being used in routine serology. I imagine they are the forerunners of similar devices now marketed by, for example, Dynatech* and Flow Laboratories†.

A more recent reference to the Cooke Autopipettor (and their automatic

*Dynatech Laboratories Inc., 900 Slaters Lane, Alexandria, Virginia 22314, USA.
†Flow Laboratories Inc., PO Box 2226, 1710 Chapman Avenue, Rockville, Maryland 20852, USA.

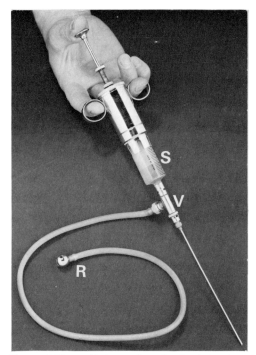

Fig. 33. A Cornwall pipette. S: spring-loaded syringe. V: one-way valve. R: tube which is placed in reservoir.

dilutor) was made by Darbyshire (1973) in his work estimating protozoan populations. He simply suggests that they produce satisfactory results in this application.

Dispensing/diluting fluids

Simple devices

One may attach a one-way valve to some of the spring-loaded syringes discussed earlier, so that one may repeatedly dispense fixed volumes of reagents. One tube leading to the valve is placed in the reservoir, and the syringe plunger is operated to expel the fluid in the syringe. As the plunger is released the syringe is filled through this valve; the process may be repeated as often as required. This is usually known as the Cornwall automatic pipette, and a wide range of such devices for dispensing volumes of the order of 250 μl and above is commercially available. A photograph of one of these dispensers is shown in Fig. 33.

Kreider and Lutz (1969) found that the use of disposable plastic syringes greatly simplified the maintenance of these devices. They found that with the particular pipette used (Becton Dickinson *) it was necessary to use a larger syringe than the nominal equivalent size syringe for a given housing, to get a

*Becton Dickinson, Rutherford, New Jersey 07070, USA.

close fit; they found a 10 ml plastic syringe fitted into the 5 ml pipette housing and a 20 ml syringe fitted the 10 ml housing. They simply stated that no pipetting accuracy is lost when using plastic syringes but they did not tell us what that accuracy is. They also told us that the use of the disposable syringes in the Cornwall pipette results in a smoother and more reliable operation and that it reduces the labour required in cleaning and sterilizing.

It is possible to automatically actuate the piston of the syringe, and a wide range of electrically operated equipment in which this is done is available. The Hook & Tucker* (Fig. 34a) and the Fisons† automatic dispensers are but two examples. They normally repeatedly dispense volumes of the order of

(a)

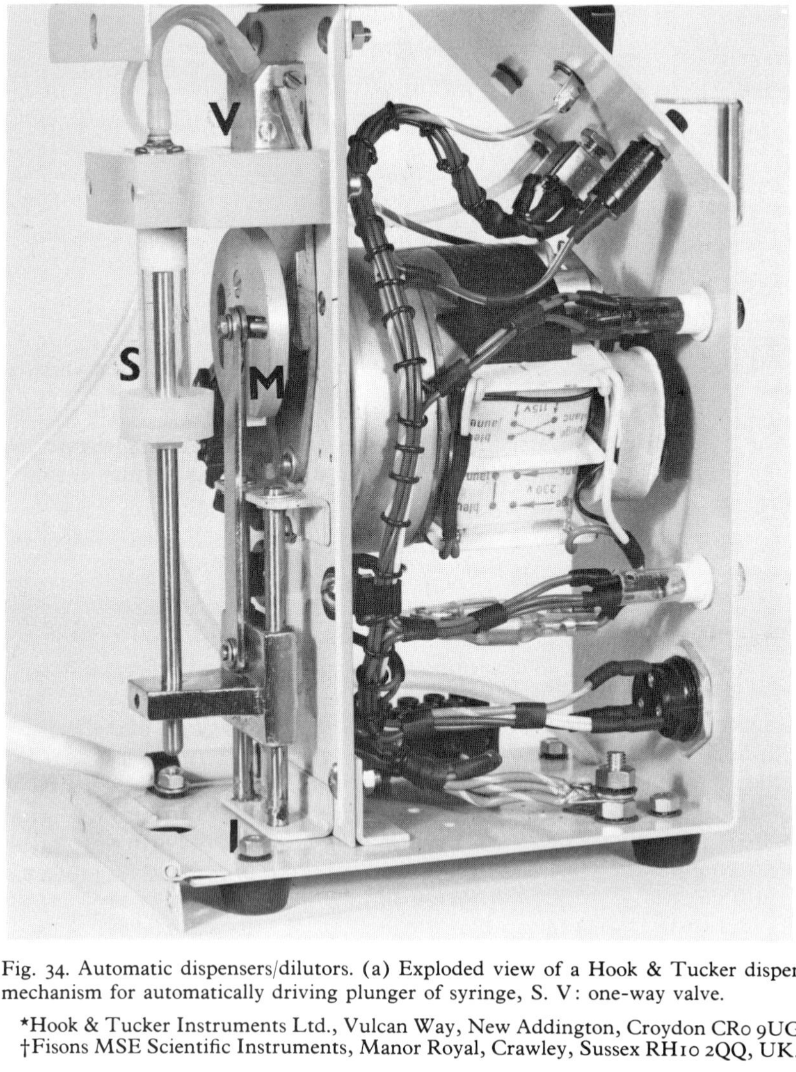

Fig. 34. Automatic dispensers/dilutors. (a) Exploded view of a Hook & Tucker dispenser. M: mechanism for automatically driving plunger of syringe, S. V: one-way valve.

*Hook & Tucker Instruments Ltd., Vulcan Way, New Addington, Croydon CR0 9UG, UK.
†Fisons MSE Scientific Instruments, Manor Royal, Crawley, Sussex RH10 2QQ, UK.

(b)

Fig. 34 contd. (b) The Fisons dispenser (right module) and dilutor (left module). (Photographs courtesy of Hook & Tucker Instruments Ltd. and Fisons MSE Scientific Instruments Ltd.)

1–10 ml using either 5, 10 or 20 ml syringes. The degree of sophistication of these devices varies; so also, therefore, does the precision with which they repeatedly dispense the nominal volume.

I do not propose to elaborate on any of these devices because there is considerable literature readily available. The Association of Clinical Biochemists produced a scientific report on automatic dispensing pipettes (Broughton et al., 1967). Although this survey was carried out some years ago and the situation has changed in the intervening period, in that details of many of the devices have been altered so that they are more precise and reliable (and also more refined), they are basically no different today.

The manufacturers of these automatic dispensers nearly always also market automatic dilutors (Fig. 34b). In these devices there are two syringes. One is used to suck up, usually into a hand-held probe, a fixed volume of sample and the other to suck up into its barrel, through a valve, a fixed volume of diluent from a reservoir. Both sample and diluent are discharged in turn through the hand-held probe. By varying the stroke length of both syringes, the ratio of sample to diluent may be varied; the ease with which this may be done varies according to the design of equipment. If the volume of diluent is high compared with the volume of sample, carry-over may be reduced to acceptable levels; however, one is advised to invariably test this for oneself.

One must remember that these dispensers and dilutors have complicated valve mechanisms which cannot normally be readily autoclaved; but some may be sterilized chemically. As a general guide, one may take it that they perform their dispensing and their diluting functions to an accuracy of the order of $\pm 1\%$ when handling volumes of the order of 1 ml or above.

More complex devices

In the electrically operated dispensing and diluting devices described above, adjustment of the stroke is often not a simple matter. It is usually done by adjusting a screw-threaded stop, and the volume has to be found by trial and error. A much more sophisticated automatic dispensing machine, but one in which the stroke length of the piston of a syringe, and thus the volume dispensed, can be simply varied, was devised by Draper and Mencken (1964).

In this device (Fig. 35), the piston of the syringe (A) is connected to a drive rod (not shown in detail) which, in turn, is connected to a cam mechanism (M, W) which actuates the rod. The stroke length of the rod, which stroke determines the volume dispensed, is varied by means of two sets of stops (I), each set being mounted round the circumference of a brass disc (J). One disc holds those stops which determine the length of the stroke by tenths of a millimetre and the other disc contains those stops that control the

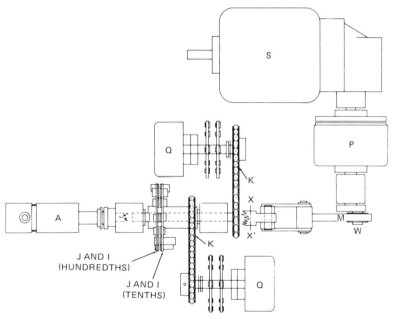

Fig. 35. Plan view of dispensing apparatus devised by Draper and Mencken. A: syringe barrel attached to fluid inlet and outlet valves. J: spacer discs supporting stops, I, which determine the stroke length of the drive rod (detail not shown). Q: rotary stepping switches. K: spacer disc drives. M, W: eccentric drive rod (and thus syringe piston) operating mechanism. S: motor. P: clutch. (From Draper, L. R. and Mencken, C. R. (1964). An automatic pipetting machine. *J. Lab. clin. Med.* **63**, 325–331.) (Drawing courtesy of Prof L. R. Draper.)

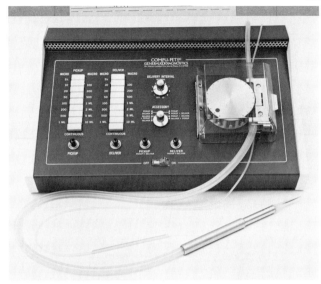

Fig. 36. The Compu-pet dispensing diluting system. (Photograph courtesy of W. R. Warner & Co.)

stroke length by hundredths of a millimetre. There are a total of 110 possible volume settings, from 0.01 to 1.09 ml in 0.01 ml steps.

The volume to be delivered is selected by pressing buttons. These in turn operate rotary stepping switches (Q) which rotate the discs through drives (K) so that the appropriate stops are in position.

The equipment was tested by determining the volume delivered in each of 12 separate samples for each of 19 different volume settings; distilled water was the test fluid. Examples of the results are: with a nominal volume setting of 1 ml the mean volume delivered was 1.0063 ml with a standard deviation of 0.0042; with a nominal volume setting of 0.1 ml the mean volume delivered was 0.0997 ml with a standard deviation of 0.0009; with a 0.01 ml nominal volume the mean volume delivered was 0.0125 ml with a standard deviation of 0.0016.

A modern version of this device is the Compu-pet dispenser/dilutor system (Fig. 36) marketed by General Diagnostics.* This is based on a multichannel peristaltic pump, which operates a 'small' bore or a 'large' bore tube and which is driven by a stepping motor, a device in which the shaft rotates a small angle (of the order of a degree) each time the motor is energized: repetitive energization is arranged electronically. With the small tube in situ, volumes of 5, 10, 20, 50, 100, 200, 500 and 1,000 μl may be sucked up and discharged; with the large tube, volumes of ten times those amounts may be handled. The particular volume required is selected merely by pressing buttons which, in turn, select the number of times the motor is energized. Similarly, one may repeatedly dispense volumes of reagents.

*General Diagnostics Division, Warner Lambert International, 201 Tabour Road, Morris Plains, New Jersey, USA.

Fig. 37. The latest model of the multiple pipetting/dispensing machine designed by Sequeira. (Photograph courtesy of Dr P. L. J. Sequeira.)

This is a very versatile machine in that it can be used for the multiple dispensing of small quantities of fluids, for multiple pipetting and diluting applications and also for serial diluting. It is reputed to be accurate and very easy to handle, although I do not doubt considerable care has to be taken when handling volumes as small as 5 and 10 μl, as, for example, movement of the tip of the tube is bound to alter the position of the fluid in the tubes and, thus, the pipetting and dispensing accuracy. I imagine it will not be long before a disposable tip is devised for the pipette in order that it can be used for pipetting infected material in exactly the same way as the disposable tip pipettes. Since the tubing is autoclavable the device may be used for repeatedly dispensing sterile fluids and cultures. In any case there are bound to be applications in which this is much more suitable and possibly more accurate than the semi-automatic pipettes, such as the Eppendorf.* A resumeé of some microbiological applications of the apparatus was published by Cremer et al. (1975). It is clearly a very valuable machine.

It was obvious that the Auto-Analyser, so widely used in chemical laboratories, would be introduced into microbiology or at least serology, and it has been used for performing the Wassermann reaction (Pugh and Gaze, 1965) and other complement fixation procedures (Vargues, 1965). It was found, however, that the washing of the apparatus by the standard volume of washing fluid was insufficient to remove all traces of antibody after the passage of a sample of serum with high antibody titre. There was some carry-over to the next specimen, and it seems that the carry-over problem could present an obstacle to the use of the continuous flow system in diagnostic serology (Taylor et al., 1968). However, Khoury et al. (1973) used the Auto-Analyser for analysing 1,000 human serum samples for *Salmonella typhi, S. paratyphi* and *Brucella abortus* by the sera agglutination procedure; the results obtained were compared with those obtained by a manual method, and the results agreed to at least 97%. Furthermore, Johnston et al. (1976) used the Auto-Analyser in a bioluminescence technique for detecting bacteriuria and found it satisfactory.

A range of equipment for performing discrete analysis methods has been devised. Sequeira (1964) developed a hand-operated apparatus for performing serological titrations (Fig. 37). In principle, the apparatus consisted of a table containing racks of reservoirs of reagents and test tubes, which could be moved so that any row of test tubes or reservoirs could be placed under a row of 12 plastic pipettes. The pipettes were lowered into the test tubes or reservoirs. Each pipette was permanently connected to a syringe and the 12 syringe pistons were operated simultaneously by a single lever to suck up and expel fluid. The volume dispensed depended on the position to which the lever was turned and, therefore, on the skill of the operator. Each syringe piston was connected directly to the lever, and the stroke length of each piston, and hence the volume dispensed, was not independently variable. The accuracy of dispensing depended, therefore, on the tolerances of the syringes as well as on the operator.

---

*Eppendorfer Gerätebau, Netherler & Hinz GmbH, 2000 Hamburg 63, Postfach 630324, West Germany.

It was necessary for the operator to know the test procedures because he controlled the order in which the agents were dispensed, the volumes dispensed and the row of tubes into which the reagents were placed. The apparatus was, however, very flexible in that variations of the order in which the reagents were dispensed, the volumes dispensed and the rack of tubes into which the reagents were placed could be introduced without difficulty. (These observations also apply to the Compu-pet and to many of the devices discussed hitherto.)

Sequeira's apparatus was very similar to that produced by Baron et al. in 1961, who claimed to be able to sterilize the pipettes by flushing them six times in boiling demineralized water; they did not discuss the technique. Sequeira's apparatus was also a simplified modification of Weitz's machine (1957). The only essential difference between Sequeira's machine and Weitz's machine was that in the latter the volume each pipette/syringe system dispensed was independently variable, even though a single lever operated them all simultaneously and, therefore, inaccuracies due to tolerances in the apparatus could be reduced at the cost of making it very complicated.

The last three devices mentioned use much larger volumes of reagents than are used in micro techniques and they have, therefore, been largely superseded.

There is little more one can usefully say. There must be hundreds of devices for distributing chemically clean fluids available today, and laboratory workers will have no difficulty in locating and acquiring one for evaluation: accuracy and repeatability need to be carefully checked.

I will conclude by repeating the main conclusions reached in Chapter 1. Beware when using this equipment, especially the most simple. Numerous helpful tips have been given on using the equipment and most have been mentioned above. For example, hold a dropping pipette vertically and look out for consistent bias in a diluting loop. If one thinks about such things and takes steps to obviate errors that may arise, all these devices can be valuable tools. Otherwise, of course, one can very easily assume all is well when in fact it is not. Constant checking of performance is the order of the day and it is always wise to do that before purchasing equipment as well as subsequently!

References

Pipetting/transferring small volumes of fluid
Diluting loops

Ashcroft, J., Platt, G. S. and Maidment, B. J. (1971). The accuracy of the microtitre technique. *Med. Lab. Technol.* **28**, 129.

Fitzgerald, S. C., Fuccillo, D. A., Moder, F. and Sever, J. L. (1974). Utilization of a further miniaturized serological microtechnique. *Appl. Microbiol.* **27**, 440.

Fuccillo, D. A., Moder, F. L., Catalano Jnr., L. W., Vincent, M. M. and Sever, J. L. (1970). *Herpesvirus hominis* types I and II: a specific microindirect hemagglutination test (34554). *Proc. Soc. exp. Biol. Med.* **133**, 735.

MacLowry, J. D. and Marsh, H. H. (1968). Semi-automatic microtechnique for serial dilution–antibiotic sensitivity testing in the clinical laboratory. *J. Lab. clin. Med.* **72**, 685.

Oberhofer, T. R. and Hajkowski, R. (1970). Evaluation of the microtiter technic for heterophilic antibody determinations. *Amer. J. clin. Path.* **53**, 498.

O'Brien, T. C., Rastogi, S. and Tauraso, N. M. (1971). Statistical evaluation of diluents and automatic diluting and pipetting machines in influenza serology. *Appl. Microbiol.* **21**, 311.
Scott, L. V., Felton, F. G. and Barney, J. A. (1957). Haemagglutination with Herpes simplex virus. *J. Immunol.* **78**, 211.
Sever, J. L., Ley, A. C., Wolman, F., Caplan, B. M., Crockett, P. W. and Turner, H. C. (1964). Utilization of disposable plastic plates with a serologic microtechnic. *Amer. J. clin. Path.* **41**, 167.

Pipettes
Ashcroft, J., Platt, G. S. and Maidment, B. J. (1971). The accuracy of the microtitre technique. *Med. Lab. Technol.* **28**, 129.
Reinisch, E. and Hall, L. (1966). A rapid microtechnic applied to the heterophile antibody test for the detection of infectious mononucleosis. *Amer. J. clin. Path.* **45**, 755.

Semi-automatic pipettes and similar devices
Kropp, B. N. (1963). An apparatus for the injection of small quantities of fluid. *Science Tools* **10**, 13.
Wederkinch, W. F. (1964). An improved micrometer syringe burette. *Scand. J. clin. Lab. Invest.* **16**, 473.
Williams, J. E. and Whittemore, A. D. (1971). Serological diagnosis of pullorum disease with the microagglutination system. *Appl. Microbiol.* **21**, 394.

Multiple pipettors
Darbyshire, J. F. (1973). The estimation of soil protozoan populations. In: *Sampling — Microbiological Monitoring of Environments.* Eds. R. G. Board and D. W. Lovelock. Academic Press, New York and London.
O'Brien, T. C., Rastogi, S. and Tauraso, N. M. (1971). Statistical evaluation of diluents and automatic diluting and pipetting machines in influenza serology. *Appl. Microbiol.* **21**, 311.
Sanderson, C. J. (1968). A multiple dropper for use with serological microtitre apparatus. *Lab. Pract.* **17**, 60.
Sanderson, C. J. (1970). A multiple dropper for micro-titration apparatus. In: *Automation, Mechanization and Data Handling in Microbiology.* Eds. A. Baillie and R. J. Gilbert. Academic Press, New York and London.
Williams, J. E. and Whittemore, A. D. (1971). Serological diagnosis of pullorum disease with the microagglutination system. *Appl. Microbiol.* **21**, 394.

Dispensing/diluting fluids
Simple devices
Broughton, P. M. G., Gowenlock, H. H., Widdowison, G. M. and Ahlquist, K. A. (1967). Automatic dispensing pipettes: an assessment of 35 commercial instruments. Assn. Clin. Biochem., Scientific Report No. 3, 1967.
Kreider, J. W. and Lutz, S. (1969). Use of disposable syringes in the Cornwall automatic pipette. *Amer. J. clin. Path.* **51**, 816.

More complex devices
Baron, S., Burch, B. L. and Uhlendorf, B. W. (1961). An automatic multiple diluting machine. *Amer. J. clin. Path.* **36**, 555.
Cremer, A. W., Mellars, B. and Stokes, E. J. (1975). The Compu-pet 100: a versatile dispenser-diluter for the mechanisation of microbiological techniques. *J. clin. Path.* **28**, 37.
Draper, L. R. and Mencken, C. R. (1964). An automatic pipetting machine. *J. Lab. clin. Med.* **63**, 325.
Johnston, H. H., Mitchell, C. J. and Curtis, G. D. W. (1976). Automation in clinical microbiology; a system for urine specimens. In: *Proceedings 2nd International Symposium on Rapid Methods and Automation in Microbiology.* Eds. H. H. Johnston and S. W. B. Newsom. Learned Information (Europe) Ltd., Oxford.
Khoury, A., Petrow, S. and Kasatiya, S. S. (1973). Automated seroagglutination test with brucella and salmonella suspensions. *Amer. J. clin. Path.* **60**, 467.
Pugh, V. W. and Gaze, R. W. T. (1965). An automated screening method for the Wassermann reaction. *Brit. J. vener. Dis.* **41**, 221.
Sequeira, P. J. L. (1964). Personal communication.
Taylor, C. E. D., Kershaw, J. W. and Heimer, G. V. (1968). An evaluation of the Technicon AutoAnalyser for automating complement-fixation tests. *J. clin. Path.* **21**, 521.
Vargues, R. (1965). The use of the AutoAnalyser for the automatic titration of antigenic preparations by means of complement-fixation. *Ann. N.Y. Acad. Sci.* **130**, 819.
Weitz, B. (1957). An automatic dispenser for multiple serological titrations. *J. clin. Path.* **10**, 200.

74

4
Devices for spreading
a culture over solid media

Small, discrete, agar plate spreaders

Spreading a culture over an agar plate, for antibiotic sensitivity testing by the disc method or for isolation and identification of organisms, is a procedure which takes up a substantial proportion of the working day in most laboratories. In consequence, an automatic device for performing this procedure would be an extremely useful piece of apparatus.

One of the first attempts to mechanize it was by Fennel et al. (1941). They used the turntable from a phonograph, which was mechanically operated, to rotate Petri dish agar plates for the inoculation of stool suspensions.

Williams and Bambury (1968) used an electrically driven turntable, with a raised lip around the circumference in order to contain the Petri dish. The dish is placed on the turntable and a bacteriological wire loop, loaded by touching a colony, is drawn radially to the centre of the plate and lifted off; thus a spiral track is traced by the loop. A sawb may be plated out directly, by moving it along the radius of the rotating dish, in exactly the same way as the loop is moved, and simultaneously rolling it. The distance between the coils of the spiral is dependent on the speed with which the loop (or swab)

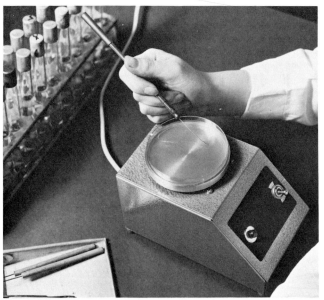

Fig. 38. The rotary turntable for spreading culture over an agar plate devised by Williams and Bambury. (Photograph courtesy of Dr R. F. Williams.)

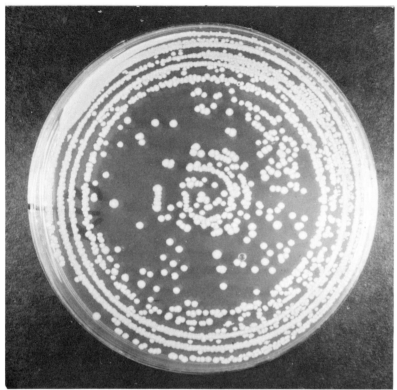

Fig. 39. A culture of *Staph. aureus* spread in the device shown in Fig. 38. (Photographic slide courtesy of Dr R. F. Williams.)

traverses the radius. The apparatus is shown in Fig. 38 and a culture in Fig. 39. It can be seen that the technique gives sufficient dilution of the culture for isolated colonies of the organism to appear after incubation.

Clearly a considerable amount of skill is required in using this type of device, as, for example, too heavy an inoculum will produce no colony isolation. No doubt experience enables the operator to reduce problems of this sort. However, there is a tendency ιo produce solid, confluent growth when the plates are inoculated from colonies and the authors utilized this phenomenon for routine antibiotic sensitivity testing by spreading a culture over the agar surface, by slowly moving the loop and loading it two or three times during the procedure. They then applied a multodisc in the conventional manner. A similar technique was used for the titration of bacteriophage, although they do not recommend the method for phage typing because ridges tend to be made in the surface of the agar, and this makes it difficult to read minor reactions. They also used the machine for spreading aliquots of dilutions of liquid culture for surface colony counting.

A bench centrifuge was modified so that agar plates could be rotated for inoculation with swab culture by Kunz and Moellering (1971); this was specifically designed for inoculating plates for antibiotic sensitivity testing.

The lid of an agar plate is attached to the head of the centrifuge, by means of double-sided cellophane tape, and this lid is used as the container in which the bottom of the agar plate is placed for inoculation. The centrifuge is run at a speed of between 200 and 250 rev/min. The authors warn against going faster than 500 rev/min because at that speed splashing occurs. I cannot help but feel, however, that even at 250 rev/min one would have to be extremely cautious of splashing and also of the culture plate taking off!

Clearly, these are very simple and flexible aids, yet despite the ease with which they can be obtained one sees comparatively few in use, although they are becoming relatively popular for performing antibiotic sensitivity (disc) tests; at least two firms market rotary turntables and stress bacteriological applications. This may well be because, although the tables may be rotated at a variety of speeds, it is a comparatively slow process; spreading a plate by the conventional method normally adopted in the routine laboratory almost certainly takes a technician less time than by this method.

A rather similar, but less flexible and more complicated, device has also been devised, because it seemed that a plate spreader would be really useful in a diagnostic laboratory only if it was sufficiently mechanized to allow the technician to carry on with other activities while the plates were being spread (Trotman, 1971).

In this machine, inoculated culture plates are automatically transported, in turn, from the holder in which they are placed manually to a unit in which the culture is automatically spread and, subsequently, to another holder. The plates are inoculated manually, by streaking a swab along a radius or by placing a drop of liquid culture close to the centre. The culture is spread by an electrically sterilizable bacteriological loop (Trotman and Draser, 1968), which rests gently on the agar plate at a point near the centre; the loop is sterilized by passing a current through it. As the table rotates the loop is

Fig. 40. An automatic rotary device for spreading culture over an agar plate.

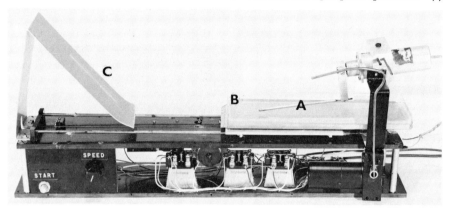

Fig. 41. The machine devised by Wilkins and his colleagues for spreading culture over an agar tray. A: standard bacteriological loop which sweeps from side to side as the tray, B, moves horizontally from right to left. C: incline of 45 degrees up which tray moves so that loop does not foul edge of the tray. (From Wilkins, Judd R. et al. (1972). Automatic surface inoculation of agar trays. *Appl. Microbiol.* **24**, 778–785.) (Photograph courtesy of Dr J. Wilkins and the American Society for Microbiology.)

automatically moved in a straight line along a radius. The rotating table together with the loop are shown in Fig. 40.

The procedure is fully automatic; the operator loads a batch of inoculated plates into one holder and the machine automatically spreads each culture in turn and places them in a second holder. The time taken to process one plate is approximately 1 min, which is slow in comparison with the manual method, but when large numbers of cultures have to be spread the time required to carry out all stages of the manual culture spreading procedure is much greater than the time required to carry out all the stages of the mechanical culture spreading procedure. The machine has been in routine use for spreading all swab and urine cultures over solid agar plates for many years. It has greatly relieved the pressure of work due to the ever increasing work-load and the shortage of technical staff.

All these devices utilize the standard 9 cm Petri dish, and, in my view, as long as cultures have to be handled, the Petri dish remains the preferred culture plate. Furthermore, since in these machines all bacteria are restricted to the spiral path, they are so much easier to read than culture plates spread manually in the conventional way.

However, a similar device, but one based on a 8.4 × 24.4 cm tray, has been devised by Wilkins et al. (1972). In this apparatus the tray is placed on a carriage which moves under a loaded inoculating loop, or a swab, mounted in a holder. This is designed in such a way that it may be raised and lowered, as appropriate, and also moved from side to side across the tray so that the loop traces a triangular curve on the agar, thus diluting the culture as before. The speed of the carriage may be varied so that the time required for it to move the total length of its traverse (approximately 23 cm) may be 105 s, 80 s or 65 s. The pressure and angle of the loop or swab relative to the plate may be altered. The tray moves up an incline of approximately 45 degrees, so that the

(a)

(b)

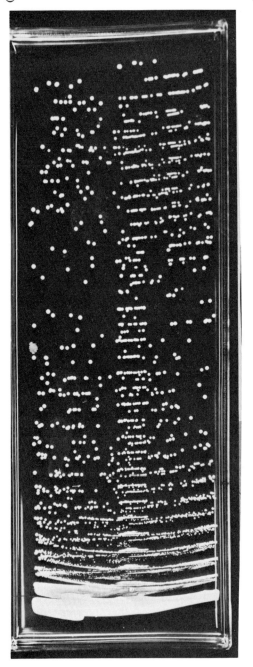

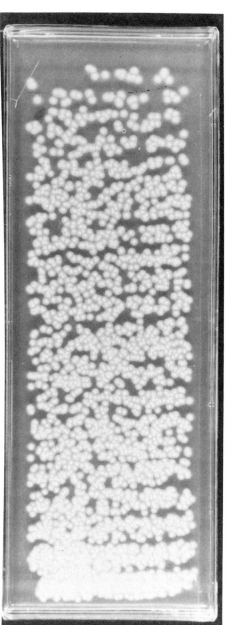

Fig. 42. Cultures spread in the machine shown in Fig. 41. (a) *Staph. epidermidis* spread with a bacteriological loop. (b) *Staph. aureus* spread with a cotton swab. (From Wilkins, Judd, P. et al.

trailing edge does not foul the loop as the tray reaches the end of its traverse. The apparatus is shown in Fig. 41.

To demonstrate its efficacy, Wilkins used a variety of organisms (*Staphylococcus aureus, Staph. epidermidis, Escherichia coli, Serratia marcescens, Pseudomonas aeruginosa, Klebsiella pneumoniae* and *Proteus mirabilis*), a 0.01 ml inoculating loop and swabs, both of which were loaded by placing them in a liquid culture of the test organism; the swabs retained approximately 0.15 ml of liquid.

As would be expected, the results obtained when using a swab culture were very different from those obtained when using a loop (Fig. 42). Single colonies can be isolated with both loop and swab but the results are highly dependent on the size of the inoculum. For the swab, the inoculum had to contain $< 10^4$ *Staph. epidermidis*/ml to obtain single colony isolation, but with the loop an inoculum of 10^9 *Staph. epidermidis*/ml gave single colony isolation.

An attractive feature of this device is that multimedia trays, with dividers to separate the media if necessary, can easily be devised so that one can produce specific trays for specific purposes. Wilkins and his colleagues used trays with two and with three different media. An example is that, in dealing with a mixture of gram-positive and gram-negative bacteria, they used a combination of blood agar to grow both organisms, azide blood agar (or phenylethyl alcohol agar) to grow gram-positive organisms whilst inhibiting gram-negative organisms, and MacConkey agar to grow gram-negative organisms whilst inhibiting gram-positive organisms.

An application of this equipment quoted was a survey of *Staph. aureus* nasal carriers. For this survey, trays containing blood agar, tellurite glycine agar and Colbeck EY agar were streaked with cotton swabs. Of the 20 persons sampled, 2 positive carriers were identified.

It is unfortunate that most of this paper was devoted to the rather idealized specimen, namely that obtained by dipping a loop or swab in a liquid culture, and only passing reference made to clinical specimens. There was one reference to a limited study of urine cultures from patients with suspected urinary tract infections. A loopful of these cultures was spread on trays containing blood agar, EMB and MacConkey agar. Good isolation of colonies was obtained. An antibiotic disc was placed on the blood agar section of the tray to allow early estimation of the susceptibility pattern. Since a 'standard' 0.01 ml loop was used, it is claimed that one can estimate the relative number of organisms in the sample.

This is an interesting development, but as indicated above I am not at all convinced that hand-operated devices such as this really do save a lot of technician time. The operator could perform the procedures more quickly by hand, and although he will have some time to spare whilst the machine is spreading the culture, this time is too short for him to do something useful. It seems, therefore, that the minimum of mechanization acceptable is a two-channel manually operated device, so that one culture is being spread automatically while the operator is preparing the next plate and loop, or swab, and loading it into the machine. In that way the minimum amount of technician time would be wasted. However, the fully automatic machine is preferable.

Other devices

From time to time one becomes involved in a discussion on the use of discrete vessels, such as Petri dishes and plastic trays, for the isolation of organisms, because there is a school of thought which believes that a continuous flexible tape, to which agar may be attached, is a more economical and better method.

One of the first of such systems was devised by Falch and Hedén (1965), for the selection of lysine-producing strains, and a more recent example is that developed by Burger and Quast (1975), specially for mass screening for epidemiological studies.

In their apparatus the agar is attached to a conventional 36 mm photographic film, and this is moved continuously under a loop which sweeps across the width of the tape perpendicular to the direction of its motion. This loop takes up the sample of liquid culture, is lowered, so as to inoculate the agar, and then sweeps back and forth across (13 mm of the width of) the tape, whilst it moves forward 60 mm. A second loop, which is situated in front of the first loop and which rests on the agar, passes through the original inoculum and the culture spread by the first loop and then starts to sweep from side to side, so as to continue the diluting process. Similarly, a third loop passes through the culture spread by the second loop, then sweeps from side to side, so that one obtains sufficient dilution of the original inoculum for single colonies to be isolated towards the end of the path traced out by the third loop. All loops are sterilized by moving them to a position to the side of the tape in which they are flamed. An example given is the spreading of *Pseudomonas aeruginosa* on Endo agar (Fig. 43a).

The authors claim that this apparatus may be used for the enumeration of bacteria and also for sensitivity testing using antibiotic discs. I am, however, a little wary of the statement that it can be used for enumeration of bacteria. My experience of using this sort of technique is that the plating is erratic, so that whilst it is more than adequate for isolating bacteria, as we have discovered whilst using our equipment for many years, it would not be suitable for enumeration purposes.

The authors refer to one serious defect. The loops are oscillated across the media at a rate of 0.5–2.5 sweep/s, but at higher frequencies droplets of the inoculum are liable to be sprayed around by the loops. Whilst the authors work at a frequency range in which they believe this phenomenon does not occur, it would seem that this is a point that needs to be constantly kept in mind with this sort of apparatus, and its performance frequently monitored very carefully. It seems to me to be a risky procedure.

A potential advantage of the continuous flexible tape system — which will become increasingly important as techniques for automatically scanning cultures for, say, automatic identification, are developed — is that a thin tape is very much easier than other configurations to scan. This was one reason why Hedén, in his Autoline system (1975), preferred to use agar mounted on discrete 1.5 × 50 cm strips of glass.

The Autoline is essentially a system made up of various modules. In the first, a variety of types of solid media are prepared. In the second, the agar is treated in a variety of ways; it may be cut up into sections, for example, and

Fig. 43. Cultures spread in the continuous tape device devised by Burger and Quast. (a) *Pseudomonas aeruginosa* on Endo agar. (b) *Proteus morganii* on blood agar. (Photographic slides courtesy of Dr H. Burger.)

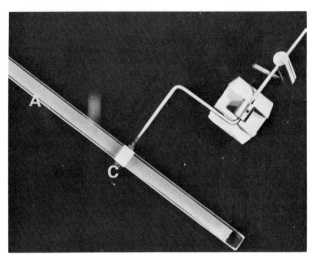

Fig. 44. The ceramic roller, C, used for inoculating a 1.5 × 50.0 cm strip of agar, A, in the Autoline system developed by Hedén. (Photograph courtesy of Prof C-G. Hedén.)

diffusion centres may be deposited in the form of paper strips. The agar may be inoculated.

For certain procedures, for example isolating organisms for metabolic pattern studies, the inoculum is spread over the agar by means of a disposable porous ceramic roller held stationary as the agar strip is moved (Fig. 44). A classical loop is moved in a straight line across the strip as it moves perpendicularly to the movement of the loop; this of course produces the same type of curve as Burger and Quast's equipment produced.

One awaits with interest the further developments of culture-spreading equipment, little of which is presently in routine use or commercially available. Whilst one may argue that in due time it will become unnecessary to isolate organisms for identification on solid media, the evidence suggests that that time is a very long way ahead; in the meanwhile there is a most valuable place for such equipment in routine bacteriology laboratories.

References

Small, discrete agar plate spreaders

Fennel, E. A., Tanaka, M. and Serizawa, S. (1941). Convenient laboratory 'gadgets'. *Amer. J. clin. Path.* **11**, 37.

Kunz, L. J. and Moellering Jr., R. C. (1971). Mechanical method of inoculating plates for antibiotic sensitivity testing. *Appl. Microbiol.* **22**, 476.

Trotman, R. E. (1971). The automatic spreading of bacterial culture over a solid agar plate. *J. appl. Bact.* **34**, 615.

Trotman, R. E. and Drasar, B. S. (1968). Electrically heated inoculating loop. *J. clin. Path.* **21**, 224.

Wilkins, J. R., Mills, S. M. and Boykin, E. H. (1972). Automatic surface inoculation of agar trays. *Appl. Microbiol.* **24**, 778.

Williams, R. F. and Bambury, J. M. (1968). Mechanical rotary device for plating out bacteria on solid medium. *J. clin. Path.* **21**, 784.

Other Devices

Burger, H. and Quast, R. (1975). Cultivating microorganisms on a substrate tape. In: *New Approaches to the Identification of Microorganisms.* Eds. C-G. Hedén and T. Illeni. John Wiley, New York and Chichester.

Falch, E. A. and Hedén, C-G. (1965). Automated selection of active strains of microorganisms. *Ann. N.Y. Acad. Sci.* **130** (Art 2), 697.

Hedén, C-G. (1975). The modular approach to the automation of microbiological routines. In: *New Approaches to the Identification of Microorganisms.* Eds. C-G. Hedén and T. Illeni. John Wiley, New York and Chichester.

5
Concluding comments

For some time I have been disappointed at the rate at which real advances are taking place in this field. I regret the exercise of writing this book has reinforced that disappointment. If one needs confirmation of the fact that progress has been slow one has only to compare the proceedings of the First International Symposium on Rapid Methods and Automation in Microbiology held in 1973 (Hedén and Illeni, 1975a, b) with those of the Second Symposium held in 1976 (Johnston and Newsom, 1976).

I find the performance of the equipment discussed generally not as good as it could and should be: it really ought not to have been necessary for me to frequently point out that it does not necessarily work in the way and with the accuracy claimed by the inventor and/or the manufacturer. I find also that the literature on this topic is generally very poor, in that often no evidence is given to justify the statements made about machine and/or microbiological test performance which is often suspect. Similarly, one is frequently left to guess how a device is made, how it works and how safe it is.

A lot of simple aids have been invented in the past 50 years, although I have been able to mention only a handful. Experience suggests that there are still very many laboratories which could make good use of some of the devices discussed, despite their limitiations. One can only hope this book will bring more awareness of the possibilities to those whose responsibility it is to organize the working of microbiological laboratories.

It is not the purpose of this book to stimulate in readers a desire to rush off and attempt to invent a new device. When one sees how much duplication of effort there has been hitherto, and the results, it becomes clear there has already been too much unco-ordinated effort devoted to such attempts. Much rationalization is called for.

On the other hand, there is a very pressing need for some professional engineers to devote their energies to devising and introducing more sophisticated automatic systems into microbiology laboratories. The more interest that is shown, and, of course, the more that automatic equipment is used, the more quickly a viable market will be created for such equipment. This is an essential prerequisite for any organization, government or quasi-government research organization or industrial organization, to invest the time and money required to improve existing and develop new automatic equipment.

This is an area in which much more effort is needed. One can only hope that such efforts are better co-ordinated than they have been hitherto. It is essential to explore all avenues, in order to increase the productivity and quality of service of microbiology laboratories, so that the service to the clinician and thus to the patient is improved.

References

Hedén, C-G. and Illeni, T. (Eds.) (1975a).
New approaches to the Identification of Microorganisms. John Wiley, New York and Chichester.
Hedén, C-G. and Illeni, T. (Eds.) (1975b).
Automation in Microbiology and Immunology. John Wiley, New York and Chichester.
Johnston, H. H. and Newsom, S. W. B. (Eds.) (1976).
Proc. 2nd International Symposium on Rapid Methods and Automation in Microbiology. Learned Information (Europe) Ltd., Oxford.

Author index

Subject index